# A View from the Divide

# A View from the Divide

## Creative Nonfiction on Health and Science

edited by Lee Gutkind

University of Pittsburgh Press

Issue 11 of *Creative Nonfiction.*

Published by the University of Pittsburgh Press, Pittsburgh, Pa.15261

Manufactured in the United States of America
Printed on acid-free paper
10 9 8 7 6 5 4 3 2 1

Library of Congress Cataloging-in-Publication data is available from the Library of Congress.

Subscriptions for individuals are $22.50 for three issues (includes one double issue) or $39 for six issues (includes two double issues). Subscriptions for libraries are $30 for three issues and $60 for six issues. Canadian subscriptions are $30 for three issues and $45 for six issues; other foreign subscriptions are $35 for three issues or $70 for six issues; both are payable in U.S. funds. Single copy price is $10, double issues, $15.95. Postmaster: Please send address changes to the Creative Nonfiction Foundation at the address listed below. Address correspondence, unsolicited material, subscription orders, and other queries to the Creative Nonfiction Foundation, 5501 Walnut St., Suite 202, Pittsburgh, Pa. 15232. Telephone: 412-688-0304; Fax: 412-683-9173; e-mail: crn2+@pitt.edu; Internet: http://www.goucher.edu/cnf. Manuscripts will not be returned unless accompanied by a self-addressed, stamped envelope.

*Special thanks to the Bayer Corporation for endowing*
*the Bayer Creative Nonfiction Science Writing Prize*
*and its generous support of this issue of* Creative Nonfiction.

*The Creative Nonfiction Foundation*
*also gratefully acknowledges*
*the Juliet Lea Hillman Simonds Foundation, Inc.*
*for its ongoing support.*

*Special thanks to the editorial board of* Creative Nonfiction,
*Lea Simonds, Laurie Graham, and Patricia Park,*
*managing editor Leslie Boltax Aizenman, and*
*the editorial staff, Christopher Hertz, Rebecca Skloot,*
*and Kathleen Veslany.*

# Contents

# A View from the Divide

# Introduction

## Doctors and Writers

LEE GUTKIND

One morning some years ago I found myself in the office of a dermatologist who, while tearing into my plantar's wart on my right foot, glanced nervously up at my chest. "Wait!" she murmured, "Melanoma!" At the time, I did not know precisely what melanoma was, but I knew the word to which it was most associated: Cancer. She tenderly touched the mole she had spotted as the likely suspect and commented: "I don't think this is malignant, but you need to have it removed immediately." She paused and continued in a hushed voice. "Not that I want to worry you." I braced myself for what was coming next. "But, three weeks from now, in a worst-case scenario, you could be dead."

I smiled bravely. "I thought you didn't want to worry me."

She did not smile back. "You need to see a surgeon."

"Well I don't know a surgeon," I said. "Who's the best in Pittsburgh?"

She replied immediately with a name that, for purposes of this essay, I have changed: "Sidney Schwartz."

"Can you make the arrangements?"

"I'll take care of everything, Mr. Gutkind."

I did not like the way she suddenly called me by my last name. Through all of our associations related to my plantar's wart, we had been on a first-name basis. Now that I was three weeks away from a painful demise, she had immediately adopted what I have come to call the "doctor's distance declaration," which establishes a direct line of withdrawal from patient interaction in proportion to severity of illness and prognosis for recovery. The more serious and potentially fatal the malady, the more physicians will study your chart and contemplate their geeky shoes, tending to walk backwards whenever the patient or family members attempt to talk with them.

Sidney Schwartz was my prototype of this syndrome. First, his nurse said that he did not need to meet me. Such a minor procedure required no personal contact or preliminary assessment. Second, it would be ten days before I could be squeezed into his schedule. And this, of course, was a favor performed on behalf of the referring dermatologist, a long-time colleague. Otherwise, it would have been a month. I appreciated the consideration. However, having done a little research about melanoma and learning how quickly it might spread, the ten days waiting with the dark specter of death I now perceived hanging over my head was one of the most anxious periods of my life. Those ten days were nothing in sheer terror to the day of the surgery.

I arrived at the Outpatient Surgical Center (OSC) at 7 A.M. for my pre-op examination. I was weighed. My pulse was taken. The necessary forms were filled out, my clothes and personal items stored away safely in a locker at the other end of the unit. I put on one of those paper hospital robes, a miniskirt model that hung about three inches above my knees, along with paper hospital booties. A nurse led me to a tiny windowless room, invited me to make myself comfortable. Dr. Schwartz was due at 8 A.M. and we were a little early.

I was immediately bored and jumpy as I waited. I had hardly been able to concentrate throughout the entire ordeal, and now was so focused on the impending surgery that I hadn't even thought to bring anything to read. I knew that Dr. Schwartz, the best surgeon in Pittsburgh, would be arriving any minute to interrupt whatever reading or work I might be doing, anyway. Even at 9 A.M., when there was no sign of Dr. Schwartz or any other doctor who might be coming to check in with me, I was confident that it was only a matter of time. It had been a little naive of me to have assumed that surgeons would be any more punctual about surgery than other physicians were in keeping appointments. After all, there were sick people needing emergency surgery throughout the city, any one of whom could be bleeding to death on the operating table with a heroic Sidney Schwartz laboring to save him. In the back of his mind, Schwartz knew that I was waiting at the OSC—and he would be rushing in here any second, breathless, spouting apologies while sharpening his scalpel, getting down to business.

I continued to believe that until around 11 A.M. when one of the nurses came in to say that she had initially been told that Sidney Schwartz had been sidetracked by an emergency procedure, but, she confessed, she was no longer certain that that was true. He wasn't answering his page and no one, including his partners, could find him. She was more embarrassed than apologetic—and she was whispering, as if confiding a dark secret or committing a crime which, in this doctor-friendly milieu, she was. Around lunchtime, the nurse reported that someone had seen Sidney Schwartz in the hospital in surgical scrubs and heading this way, but when he did not show up by 1 P.M. she contacted the operator who began to page him over the hospital intercom. There were thirty separate pages over the next hour and a half—I counted every one of them as I sat, a prisoner in that windowless waiting room, listening to the air-conditioning fan and the muffled activity in the hallway behind my closed door. The hollow, persistent sound of the paging operator summoning Schwartz to the Outpatient Surgical Center triggered within me a new and more acute wave of anxiety and fear. Periodically, I wandered out of my room and inquired at the nurse's station for an update. Twice I used the nurse's phone to call Schwartz's office to complain, but after being frozen in "hold" limbo for many minutes as Schwartz's secretaries were attempting to locate him, I hung up the phone and retreated back to my cell.

At 3 P.M., when the nurse came in carrying my clothes to apologetically announce that her shift was over and that the entire unit was closing down for the day, I was a complete wreck. To me, this was an omen, a clear message that the melanoma was malignant and that I was going to die. After all, the dermatologist had said "three weeks"; nearly half of the last days of my life had been wasted waiting for Sidney Schwartz, the best surgeon in Pittsburgh, to stand me up. This was torture: eight hours in a windowless waiting room, no television, radio, or reading material. Even convicted murderers were given breakfast, lunch, perhaps an hour-long walk in the yard for fresh air—and real clothes. You could hardly take a walk in the yard in a paper miniskirt and pastel blue booties.

In retrospect, it was a mistake to have arrived at the OSC without a friend to keep me company or demand help, but I was recently divorced and feeling as if I needed to learn to confront the challenges of life on my own without leaning on a partner. Later, collapsing at home in my bed, I listened to the messages on my answering machine. The last one was from a secretary in Schwartz's office explaining that in the frenetic rush of his day, Dr. Schwartz had simply overlooked my procedure and that he would be pleased to reschedule for the following week. She offered a couple of dates and times and suggested that I return her call if I was interested.

But my one and only interest was in killing him if, of course, I lived through the melanoma. That fear had now considerably paled in comparison to my burning hatred of Sidney Schwartz. I wrote a number of letters, never sent, and composed a slew of speeches, never delivered, over a long period of years. Not that I was afraid of Sidney Schwartz or too sick from chemotherapy (which, in fact, never happened—I found another surgeon who removed the mole a few days later, which, after a quick biopsy, was judged benign). The truth is, I never confronted Dr. Schwartz. I was too enraged to simply yell and scream and bash his head in; rather, I wanted to humiliate Sidney Schwartz in front of his colleagues, friends, and family—the people who most respected him. My dream was that I would serendipitously come into contact with him one day, face-to-face, preferably at a dinner party. He wouldn't know me, but I would know him, and I would charm and befriend him, along with everyone else at the table, all the while gently guiding the conversation toward the issues of ethics and morality in medicine—the Hippocratic Oath, the physician's responsibility to the patient, all the good stuff about which physicians love to expound, at which point I would begin my melanoma story in basically the same way I have started the story here, by describing the dermatologist, and the hour-by-hour, soul-twisting torture in the windowless room waiting for the surgeon to appear. But I wouldn't say the surgeon's name—not for a while.

I'd wait for the precise and proper moment, skillfully unfolding the details of the story and building suspense and curiosity about the identity of the surgeon, dropping a few subtle, pointed clues. In this scenario, I can literally picture all of the people at the long dinner table, spellbound, heads strained forward to listen as the scenes and incidents I relate gallop toward the conclusion. And I can also picture the object of my anger, Sidney Schwartz, growing ever more uncomfortable in his chair as the intimate nature of my story finally begins to touch some inner chord of awareness. I know in my heart that he is beginning to suspect the humiliating reality that is approaching him. At the end of the story, with my dinner companions completely primed and totally empathetic and with my final words of dismay echoing through the quiet, darkening dining room, someone breaks the silence and asks: "Who was this horrible doctor?"

At which point, I carefully place the coffee cup in its saucer and turn ever so slowly toward my nemesis, my enemy, the object of my pent-up, bone-scraping rage, Sidney Schwartz, and look him straight in the eye, as I quietly announce: "He is sitting among us at this table."

It was this dream of revenge through the consummate humiliation of Sidney Schwartz specifically and my melanoma experience generally that led me eventually to write about medicine and science.

My first books were about my two great passions: First, motorcycles and the subculture that surrounds them, for which I traveled extensively, cruising most of the lower forty-eight over a period of three years. The second was baseball. I spent one season shadowing a crew of National League umpires from ballpark to ballpark. But being forgotten and abandoned by Sidney Schwartz, the person to whom in good faith I had blindly entrusted my life, began a process of evaluation that was to change my entire artistic orientation for the next fifteen years. What kind of people could devise a system in a world with the Hippocratic Oath as the bedrock of healing, that would engender such an impersonal and unresponsive atmosphere, I wondered? Why is a medical center designed to function primarily for its doctors, rather than for the patients it is supposedly created to treat and serve? These were the questions that came to intrigue and drive me.

Since then, I have written four books about the humanistic aspects of the high-tech medical world. The researching technique I use for these books can be called "total immersion," meaning that I literally move into the medical setting in which I have interest and invest months and sometimes years learning everything possible about aspects of modern medical technology from the points of view of all of the actors—physicians, nurses, patients, family members—anyone with a significant role in the subject that I am investigating. Using scene, dialogue, and specific detail, I hope to capture and relate the dramatic stories of people enmeshed in real life incidents and challenges—the defining moments of my immersion experiences—and the universal meanings behind what I see and re-create.

For my first book about the medical world, *Many Sleepless Nights,* an examination of the world of organ transplantation, for example, I was initially enticed by the potential of transplantation to save lives. What began to affect me, though, was the pain and suffering organ transplantation (and other forms of high-tech medicine) caused, not only to patients, but to the families who had to support them through surgery, recovery—and far beyond. But what also struck me then and continues to affect me now, years later, is the dearth of humanity at many points along the process. Not only does the technology (medications, machinery, computerization) divide the patient and doctor, but it also serves, seemingly, to deny the necessity for or expectations of courtesy and compassion.

On the organ transplant service, I once listened to a prominent surgeon impatiently interrupt a resident who was carefully explaining a procedure to a family member, prompting him to "save lives first—answer questions later." Another surgeon told me, in defense of his insensitive behavior, "Psychologic trauma and all that stuff is important, but it doesn't make a goddamn difference if you are well-adjusted and dead." Saving lives of dying patients becomes a surgeon's obsession, but in the process, such a single-minded and narrow pursuit seems to alter or destroy their sense of purpose—the reasons surgeons endured years of medical training and took the Hippocratic Oath.

I never experienced that feeling with veterinarians, however, for my most recent book, *An Unspoken Art*. Here were men and women, obviously interested in lifesaving, but dedicated and devoted not only to the physical, but also the psychological well-being of their patients. This is the ironic missing link in human medicine: many of our caregivers do not regard or treat their patients as human beings, as I initially discovered so many years ago with Sidney Schwartz. And this is the part of veterinary medicine I will never forget, that part that includes philosopher Erich Fromm's observation that humankind "is biologically endowed with the capacity for biophilia, the passionate love of life and all that is alive"—and the way in which it is exemplified in veterinary medicine—by human contact: touching.

I was lucky to witness a revolutionary cryosurgical technique at the University of Pennsylvania on a prized harness racer named Cam Fella. The procedure was fascinating and exciting, but the most memorable part of the experience occurred long after the surgery, long after the owners had departed and most of the entourage and the curious onlookers had disappeared. Eight exhausted veterinarians and nurses, all women, remained in the recovery area with Cam Fella, sitting in a circle, elbow to elbow, keeping him calm. Touching him. Kissing him. Talking to him. Until he was awake enough to stand on his own and navigate the winding path back to his stall.

Although I connected with the veterinarians in a special way because of their compassion, I came to admire and respect the overall perseverance of the men and women in the medical community of all specialties and orientations—especially organ transplantation—relentless in their drive to solve scientific puzzles in the face of ongoing and debilitating defeat, not over a period of days and weeks, but years—or even decades. I discovered a great affinity for these egocentric scientists because I recognized early on that we shared the same set of values concerning commitment to a potentially unreachable and sometimes unimaginable goal, along with the same faith in our basic vocational ethic, which was to come to work every day without any expectation, except to function at the highest level possible for as long as possible until or whenever their project or study was completed, its thesis tested, if not proven. This too is the exact orientation of all serious writers I know: to write every day, to invent or capture memorable and true characters and stories, to discover insight spontaneously and to write sentences that invigorate and charm a reader—without visions of glory or grandeur.

From the information I have gathered through subtle inquiries, Sidney Schwartz was not such a dedicated and driven scientist; rather, he was a skilled and committed surgeon, popular for his quick and efficient procedures, but overloaded and somewhat disorganized. Sidney Schwartz almost died—many years after the day he stood me up on the operating table, of hepatitis, contracted, I am told, from one of his patients. When I heard about Sidney Schwartz's disease, I have to admit to a momentary tinge of elation—not because I wanted him to die or to be in pain, but in the hope that there was some sort of prevailing justice. So that when his colleagues began to pull away from him, as they undoubtedly would, he too would have the experience of being the unempowered patient rather than the all-powerful doctor.

I doubt that the justice I imagined actually came about in Sidney Schwartz's case, but I am vengeful enough to hope that it did, especially now that I know, in retrospect, that Schwartz received a liver transplant (another irony; a few years earlier I might have observed or scrubbed-in on the procedure), survived and returned to his practice, a result that is not nearly as routine as the medical community or the media might lead an unaware consumer to believe. In fact, I am told that he recently built a new and luxurious house in my neighborhood, although I still don't know what he looks like and don't intend to find out. At this point in my life (and in his), I suspect that if I met him at a dinner party, we would get along fine. Sidney Schwartz and I would discover a special connection because I could and would empathize with the odyssey of the liver transplant experience.

Actually, I guess I could say that despite his unconscionable albeit unknowing actions, I owe Sidney Schwartz a debt of gratitude for inadvertently directing me toward a phase of my career that helped fuse what had seemed, at one time, to be diametrically opposed concepts—creativity and science. Because of Sidney Schwartz, a doctor too busy to remember a frightened patient, I have been enlightened.

~

Many of the physicians I met and observed, surgeons especially, were relentless and persistent and also scientifically daring—willing to take risks—the hallmark of both great science and great literature and one of the reasons this double issue of Creative Nonfiction exists. We believe in the symmetry between science, art and humanism, a concept that this collection will illustrate, beginning with the lead essay by Alison Hawthorne Deming that discusses the harmony between scientific language and poetics. Deming, who is director of the University of Arizona Poetry Center, is the author of a number of books, including *Science and Other Poems,* a collection that won the prestigious Walt Whitman Award of the Academy of American Poets.

Deming is also the winner of the first Bayer Creative Nonfiction Science Writing Award, sponsored by Creative Nonfiction and the Bayer Corporation for this essay, for which she receives a $500 award. We will not be sponsoring a science-oriented issue in 1999, but for the following year the Bayer Corporation has agreed to expand the award to $2,000 in honor of the millennium. To my knowledge, this is the largest cash award for a single essay offered by a literary journal—anywhere.

Although Deming is a poet who writes about science, some of the essayists collected in this double issue are scientists who write like poets. Gerald Callahan is an immunologist in the Department of Pathology at Colorado State University in Fort Collins while James Glanz, runner-up for the Bayer award, who received his Ph.D. at Princeton, is a writer at *Science* magazine. Two members of the National Aeronautics and Science Administration (NASA) are contributors to this issue. Scott A. Sandford investigates planetary systems, astrochemistry, and the origin of life. Susan Adkins had been a NASA librarian for most of her professional life, but her passions were writing and flying, which is how she is remembered in this book, as a fine essayist and dedicated pilot, who died in a flying accident a few months before this issue went to press. She was forty-nine.

# Science and Poetry

## A View from the Divide

ALISON HAWTHORNE DEMING

The most remarkable discovery made by scientists is science itself. The discovery must be compared in importance with the invention of cave-painting and writing. Like these earlier human creations, science is an attempt to control our surroundings by entering into them and understanding them from inside. And like them, science has surely made a critical step in human development and cannot be reversed.
—JACOB BRONOWSKI

The great English poet John Donne published *An Anatomy of the World* in 1611, one year after Galileo's first accounts of his work with the telescope appeared. The poem was probably commissioned as a funeral elegy for Elizabeth Drury, who died at age fourteen, the daughter of a wealthy London landowner. But that loss is not the only spiritual dislocation the poem commemorates. The universe suddenly had been peppered with ten times the stars that had been there before. The perception of the Earth's place in that expanded (though not yet expanding) universe had been thrown into metaphysical revolution. Donne was not convinced by the new theories of Copernicus and Brahe placing the sun at the center and the Earth as merely a whirling outlier, but he took them seriously enough that one can feel his inner sense reeling:

> And new philosophy calls all in doubt,
> The element of fire is quite put out;
> The sun is lost, and th' earth, and no man's wit
> Can well direct him where to look for it.
> And freely men confess that this world's spent,
> When in the planets, and the firmament
> They seek so many new . . .
> 'Tis all in pieces, all coherence gone . . .

Part of his task as a poet was to integrate this new information about the nature of reality with his beliefs and emotions, to give a voice to his very process of confusion, his struggle for equilibrium in a newly unstable world. It is difficult to imagine a conceptual change more profound than the one experienced during the first century of modern science. The Copernican Revolution meant that people could no longer trust their senses. The experience of observing the sun circle around the Earth, as one might continue to witness every day, was no longer the truth. What then could be the value of the senses, of experience, after one has learned that the truth requires tests, measurement, and collective scrutiny?

That shift in earthlings' fundamental sense of place may not seem like a big deal now. We have had a few centuries to get used to living with its psychic disjunction. But science (along with its headstrong, profiteering offspring technology) has not slowed down in presenting artists with destabilizing new realities. As we race toward the millennium, the dizzying changes in chaos and quantum and genome theories, in the neurophysics of the brain and the biotechnology of reproduction, and in the search for the Theory of Everything, can send the amateur science watcher into a state of permanent vertigo. Indeed, I am surprised at how few contemporary artists, and in particular poets, have captured that sense of reeling. Certainly there are some—A. R. Ammons, Richard Kenney, Pattiann Rogers, James Merrill, Diane Ackerman, Miroslav Holub, May Swenson, Jorie Graham, and Loren Eiseley all have made footholds in the shifting terrain.

Nevertheless, the view from either side of the disciplinary divide seems to be that poetry and science are fundamentally opposed, if not hostile to one another. Scientists are seekers of fact; poets revelers in sensation. Scientists seek a clear, verifiable, and elegant theory; contemporary poets, as critic Helen Vendler recently put it, create objects that are less and less like well-wrought urns, and more and more like the misty collisions and diffusions that take place in a cloud chamber. The popular view demonizes us both, perhaps because we serve neither the god of profit making nor the god of usefulness. Scientists are the cold-hearted dissectors of all that is beautiful; poets the lunatic heirs to pagan forces. We are made to embody the mythic split in Western civilization between the head and the heart. But none of this divided thinking rings true to my experience as a poet.

In my high school biology notebook, which I keep with the few artifacts of my youth that continue to interest me, among the drawings in meticulous colored pencil of the life cycles of diploblastic coelenterata and hermaphroditic annelids, is a simple schematic of an unspecified point in human history at which science and religion took separate paths as ways to understand the world. I still can picture my biology teacher with his waxy crewcut and a sport jacket standing at the blackboard explaining the schism as simply as if it were an intersection on a highway. What could be tested and measured took the road of science, he said, and the unknown took the other. It is the drawing I remember most keenly because it seemed to me, even then, puzzling. How could the great questions about the nature of existence be separated into subjects, professions, vocabularies that had little to say to one another? Wasn't everyone, wasn't all knowledge and ignorance, joined by the simple desire to know the physical world, to learn how "I" got to be a part of it and to make some meaning out of our collective existence? How would the world look, I wondered, if one could see it from a point prior to that split?

I was hooked. Science became for me, not the precinct of facts, but the place where the most interesting questions were asked. I knew that no matter how much the professional rigor of science demanded objectivity, there would always be the curiosity and bewilderment of a human being hiding somewhere in the data. And though decades would pass before I heard the name Heisenberg, I already began to sense what I would later read: "Even in science the object of research is no longer nature itself, but man's investigation of nature."

That year for the school science fair I conducted an experiment on white mice to see if they would get skin cancer from tobacco. I distilled the smoke of cigarettes into a vile black paste and pasted it on their pink depilatoried backs. For the control, I used a known carcinogen, benzanthracene, I believe. I kept the cages in the cellar playroom of my family's home, tucked on top of the piano. All of my subjects developed lesions. I was a smoker at the time (a fact that did not favorably impress the fair's judges). After the fair, my biology teacher, also a smoker, helped me etherize my charges. And that's about the extent of my career as a scientist—a far cry from the lofty questions that had spurred my interest. The experience led, twenty years later, to the poem "Science," in which I began to discover the mythology of science as a guiding force in our civilization, a force like that of ancient gods, capable of generating both transforming hope and abject humility, a discipline that explores both the nature of reality and the nature of ourselves.

It is the mythological significance of science that continues to attract me as a poet, not simply the guiding stories and metaphors—"The Big Bang," "The Tangled Bank," and "The Neural Jungle"—but also the questions that drive scientific endeavor, the ambiguities and uncertainties it produces. No one with a television can fail to perceive that current scientific events play a prominent role in American culture, whether we understand the events or not. The incredible staying power of *Star Trek,* in all its combinations and permutations and spinoff subculture, attests to this. Where will those wacky intergalactic science nerds lead us next? But actual science events—news of research, for example, with the Hubble space telescope, the genome mapping project, biogenetic engineering or the extinction of species—meets more than its share of the public's hostility and skepticism toward authority of any stripe. Today fewer Americans than ever believe scientists' warnings about global warming and diversity loss. Their skepticism stems, in part, from the fact that to a misleading extent the process of science does not get communicated in the media. What gets communicated is uncertainty, a necessary stage in solving complex problems, not synonymous with ignorance. But the discipline itself is called into question when a scientist tells the truth and says, in response to a journalist's prodding, "Well, we just don't know the answer to that question."

The public's skepticism stems from other sources. Everyone knows all too well that an expert can be found (and paid) to take any scientific position that will support the claim of a special (likely corporate) interest. Coupled with this, the public is generally ignorant about the most basic science concepts. In a 1995 study fewer than 10 percent of U.S. adults could describe a molecule, only 20 percent could minimally define DNA, and slightly fewer than half knew that Earth rotates around the sun once a year. Lacking basic science literacy, one is unable to assess whether or not an expert opinion is persuasive. The capacity to appreciate such tropes as "the selfish gene," "punctuated equilibrium," "the greenhouse effect," or "cascading extinctions" is beyond hope.

What science bashers fail to appreciate is that scientists, in their unflagging attraction to the unknown, *love* what they don't know. It guides and motivates their work; it keeps them up late at night; and it makes that work poetic. As Nobel Prize–winning poet Czeslaw Milosz has written, "the incessant striving of the mind to embrace the world in the infinite variety of its forms with the help of science or art is, like the pursuit of any object of desire, erotic. Eros moves both physicists and poets." Both the evolutionary biologist and the poet participate in the inherent tendency of nature to give rise to pattern and form.

In addition to the questing of science, its language also attracts me—the beautiful particularity and musicality of the vocabulary, as well as the star-factory energy with which the discipline gives birth to neologisms. I am wooed by words such as "hemolymph," "zeolite," "cryptogram," "sclera," "xenotransplant," and "endolithic," and I long to save them from the tedious syntax in which most science writing imprisons them. As a friend from across the divide has confirmed, even over there the condition of "journal-induced narcolepsy" is all too well known. The flourishing of literary science writers, including Rachel Carson, Lewis Thomas, E. O. Wilson, Oliver Sacks, James Gleick, Stephen Jay Gould, Gary Paul Nabhan, Evelyn Fox Keller, Natalie Angier, David Quammen, Stephen Hawking, Terry Tempest Williams, and Robert Michael Pyle, attests to the fruitfulness of harvesting this vocabulary, of finding means other than the professional journal for communicating the experience of doing science. I mean, in particular, those aspects of the experience that will not fit within rigorous professional constraints—for example, the personal story of what calls one to a particular kind of research, the boredom and false starts, the search for meaningful patterns within randomness and complexity, professional friendships and rivalries, the unrivaled joy of making a discovery, the necessity for metaphor and narrative in communicating a theory, and the applications and ethical ramifications of one's findings. Ethnobotanist and writer Gary Paul Nabhan, one of the most gifted of these disciplinary cross-thinkers, asserts that "narrative and metaphor are more honest, precise and comprehensive ways of explaining an animal's life history than the standard technical format of hypothesis, materials, methods, results and discussion."

Much is to be gained when scientists raid the evocative techniques of literature and when poets raid the language and mythology of scientists. The challenge for a poet is not merely to pepper the lines with spicy words and facts, but to know enough science that the concepts and vocabulary become part of the fabric of one's mind, so that in the process of composition a metaphor or paradigm from the domain of science is as likely to crop up as is one from literature or her own backyard. I subscribe to *Science News* to foster that process, not for total comprehension, but to pick up fibers and twigs, so to speak, that I might tuck into the nest of my imagination.

Here is a recent poem of mine that operates on this principle, a poem that pokes fun at some of the rather curious practices of my naturalist friends, while praising the deeper longing that motivates them:

The Naturalists

When the naturalists
see a pile of scat,
they speed toward it
as if a rare orchid
bloomed in their path.
They pick apart
the desiccated turds,
retrieving a coarse
black javelina hair
or husk of piñon nut
as if unearthing gems.
They get down on their knees
to nose into flowers
a micron wide—belly flowers,
they say, because that's
what you get down on
to see them. Biscuitroot,
buffalo gourd, cryptograms
to them are hints of
some genetic memory
fossilized in their brains,

an ancient music they try
to recall because,
although they can't quite
hear the tune, they know
if they could sing it
that even their own wild
rage and lust and death
terrors would seem
as beautiful as the
endolithic algae
that releases nitrogen
into rocks so that
junipers can milk them.

I will leave the analysis, both literary and psychological, to the critics. What pleases me about this poem (other than the fact that I managed to use both "cryptogram" and "endolithic" in a single poem) is the way that an interesting fact (that rock-dwelling algae are a major source of nutrient for junipers growing in rimrock country) becomes a metaphor for inner, meditative aspects of the naturalists' work. As Leo Kadanoff wrote, "it is an experience like no other experience I can describe, the best thing that can happen to a scientist, realizing that something that's happened in his or her mind exactly corresponds to something that happens in nature." And so it is with poets.

But science and poetry, when each discipline is practiced with integrity, use language in a fundamentally different manner. Both disciplines share the attempt to find a language for the unknown, to develop an orderly syntax to represent accurately some carefully seen aspect of the world. Both employ language in a manner more distilled than ordinary conversation. Both, at their best, use metaphor and narrative to make unexpected connections. But, as Czech immunologist and poet Miroslav Holub points out, "for the sciences, words are an auxiliary tool." Science—within the tradition of its professional literature—uses language for verification and counts on words to have a meaning so specific that they will not be colored by feelings and biases. Science uses language as if it were another form of measurement—exact, definitive, and logical. The unknown, for science, is in nature. Poetry uses language itself as the object—as Valéry said, "poems are made with words not ideas"—and counts on the imprecision of words to create accidental meanings and resonances. The unknown, for poetry, is in language. Each poem is an experiment to see if language can convey a shapely sense of the swarm of energy buzzing through the mind. The elegance and integrity of a scientific theory has to do with the exclusion of subjective, emotional factors. The elegance and integrity of a poem is created, to a great extent, by its tone, the literary term used to describe the emotional hue of a poem conveyed by the author's style. The aim of scientific communication is to present results to the reader, preferably results that could be obtained by another researcher following the same procedures; the aim of poetry is to produce a subjective experience, one that could be obtained through no other means than the unique arrangement of elements that make up the poem. Perhaps, among scientific specialties, the work of evolutionary biologists comes closest to that of poets, because its object of study (the biological past) is intangible, its method narrative: to tell the story of life on earth.

While the two disciplines employ language in different ways, they are kindred in their creative process. W. I. B. Beveridge, a British animal pathologist, has written several useful books about the mental procedures that lead to new ideas, whether in science, art, or any other imaginative enterprise. "Most discoveries that break new ground," he asserts, "are by their very nature unforeseeable." The process is not purely rational, but dependent upon chance, intuition, and imagination. He analyzes the part that chance plays by delineating three different types of discovery in which it is a vital factor: intuition from random juxtaposition of ideas, which is an entirely mental process; eureka intuition, which results from interaction of mental activity with the external world; and serendipity, which is found externally without an active mental contribution.

Random intuition links apparently unconnected ideas or information to form a new, meaningful relationship. It is like those children's books with the pages split in half. You combine a lumberjack's torso with a ballerina's legs, and—presto—a chimera is born. Eureka intuition is best represented by two classic examples. While visiting the baths, Archimedes suddenly awoke to a significant principle that would enable him to measure the volume of an object based upon the amount of water it displaced. At the time he had been wrestling with a royal problem. The ruler Hiero suspected that he had been cheated by the goldsmith who had crafted his crown. Archimedes' job was to determine the volume of the crown, so as to learn, from its weight, whether or not it had been made of pure gold. The Roman architect Vitruvius recounts the eureka moment of Archimedes' discovery:

> When he went down into the bathing pool he observed that the amount of water which flowed outside the pool was equal to the amount of his body that was immersed. Since this fact indicated the method of explaining the case, he did not linger, but moved with delight he leapt out of the pool, and going home naked, cried aloud that he had found exactly what he was seeking. For as he ran he shouted in Greek: eureka, eureka.

The second classic example is that of Isaac Newton who watched an apple fall from a tree and saw in its motion the same force that governs the moon's attraction to the earth. Eureka intuitions occur, Beveridge explains, when one "*seeks* random stimulation from outside the problem," and they "evoke the exclamation 'I have found it!'"

In serendipity one finds something one had not been looking for:

an unusual event, a curious coincidence, an unexpected result to an experiment. The term was coined by Horace Walpole in 1754 after an ancient fairy tale that told of the three princes of Serendip. "They were always making discoveries, by accident and sagacity, of things which they were not in quest of . . . you must observe that *no* discovery of a thing you *are* looking for ever comes under this description." Examples of serendipity are Columbus finding the New World when he was seeking the Orient, and Fleming discovering penicillin when mold accidentally grew on his discarded staphylococcus culture plates. For discoveries to be made by serendipity, more is required than luck. Beveridge emphasizes that "accidents *and* sagacity" are involved: One must be keenly observant, adventuresome, ready to change one's mind or one's goal.

I think of poetry as a means to study nature, as is science. Not only do many poets find their subject matter and inspiration in the natural world, but the poem's enactment is itself a study of wildness, since art is the materialization of the inner life, the truly wild territory that evolution has given us to explore. Poetry is a means to create order and form in a field unified only by chaos; it is an act of resistance against the second law of thermodynamics that says, essentially, that everything in the universe is running out of steam. And if language is central to human evolution, as many theorists hold, what better medium could be found for studying our own interior jungle? Because the medium of poetry is language, no art (or science) can get closer to embodying the uniqueness of a human consciousness. While neuroscientists studying human consciousness may feel hampered by their methodology because they never can separate the subject and object of their study, the poet works at representing both subject and object in a seamless whole and, therefore, writes a science of the mind.

I find such speculation convincing, which is probably why I became a poet and not a scientist. I could never stop violating the most basic epistemological assumption of science: that we can understand the natural world better if we become objective. Jim Armstrong, writing in a recent issue of *Orion*, puts his disagreement with this assumption and its moral implications more aggressively:

> Crudely put, a phenomenon that does not register on some instrument is not a scientific phenomenon. So if the modern corporation acts without reference to "soul," it does so guided by scientific principles—maximizing the tangibles (profit, product output) that it *can* measure, at the expense of the intangibles (beauty, spiritual connectedness, sense of place) that it cannot.

Clearly a divide separates the disciplines of science and poetry. In many respects we cannot enter one another's territory. The divide is as real as a rift separating tectonic plates or a border separating nations. But a border is both a zone of exclusion and a zone of contact where we can exchange some aspects of our difference, and, like neighboring tribes who exchange seashells and obsidian, obtain something that is lacking in our own locality.

One danger to our collective well-being is that language continues to become more specialized within professional disciplines to the extent that we become less and less able to understand one another across the many divides, and the general public becomes less and less willing to try to understand what any of the experts are saying.

Writing the Lowell lectures at Harvard in 1925, Alfred North Whitehead foresaw the dangers of specialization. In his work on the metaphysical foundations of science, *Science and the Modern World*, the mathematician cautioned that with increasing scientific and technological refinements

> the specialized functions of the community are performed better and more progressively, but the generalized direction lacks vision. The progressivism in detail only adds to the

> danger produced by the feebleness of coordination . . . in whatever sense you construe the meaning of community . . . a nation, a city, a district, an institution, a family or even an individual . . . The whole is lost in one of its aspects.

The whole that we are losing is the belief in the integrity of life. We may have confidence in the earth's fecundity, its cleverness in reinventing life even after cataclysmic extinction spasms. But we are coming to suspect that the future of humanity is a detail that is at odds with the well-being of the whole. "If present trends continue," Beveridge wrote in 1980, "only about one per cent of the Earth's surface will remain in its natural state by the turn of the century and a large proportion of the animal species will be doomed to extinction." Civilization is speeding up the process of evolution so fiercely that species counting on their genes to keep up lose ground as fast as we either claim or ruin it.

In addition to widespread species loss, the planet is experiencing widespread loss of cultures and languages. Jared Diamond, in a 1993 article, wrote that at the present rate of loss the world's six thousand modern languages could be reduced within a century or two to just a few hundred. He estimates that it takes over a million speakers for a language to be secure. The majority of languages are "little" ones having around five thousand speakers, and they are fostered by geographic isolation. The Americas at the time of the Conquest had a thousand languages; Diamond speculates that there may have been tens of thousands of languages spoken before the expansion of farmers began around eight thousand years ago. As remote regions become less remote, the little languages erode. Since each language represents not merely a vocabulary and set of syntactical rules, but a unique way of seeing the world, these losses diminish our collective heritage.

Yet one can take some heart that specialized vocabularies within the large languages are burgeoning, and in no field are they doing so with more gusto than in science, providing fresh instruments for seeing the world. And as Whitehead wrote, "a fresh instrument serves the same purpose as foreign travel; it shows things in unusual combinations. The gain is more than a mere addition; it is a transformation."

For both science and poetry the challenges lie in taking on the complexity of the most interesting questions (formal, technical, theoretical, and moral) within our fields without losing connection with people outside of our fields. The idea of poetry with which I grew up was, I suppose, a particularly American one—that is, as an escape from the burdens of community into extreme individuality, a last bastion of rugged individualism from which one could fire salvos at an ever more remote, corrupt, and inane culture. Historically, however, the voice of poetry has not always been construed to be the heightened voice of individualism. Among the original forms of humanity, art was unified with prayer and healing science. Poems and songs were manifestations of a collective voice, of spells and visions, of spirits returning from the dead. Such poetry transcended individualism, rather than celebrating it. We may have gained much in terms of technical and artistic refinement through our specialized disciplines, but we have lost the belief that we can speak a common language or sing a common healing song.

If poetry today needs anything, it needs to move away from its insular subjectivity, its disdain for politics and culture and an audience beyond its own aesthetic clique. A poem reaches completion in finding an audience. The challenge today is to reach an audience not comprised solely of members of one's own tribe. We must write across the boundaries of difference. A poet finds a voice by holding some sense of audience in mind during the process of composition. It is one of the questions most frequently asked of poets: for whom do you write? And the answers range from writing for posterity to writing for (or against) one's literary predecessors, from writing to an intimate other, to, as Charles Wright once said, writing for the better part of oneself.

I write with an inclusive sense of audience in mind, hoping to cross the boundaries that separate people from one another. I would like to write a poem that other poets would appreciate for its formal ingenuity, that scientists would appreciate for its accuracy in attending to the phenomenal world, that the woman at the check-out counter at Safeway would appreciate for its down-to-earth soul, and that I would appreciate for its honesty in examining what troubles and moves me.

The great biology watcher Lewis Thomas once raised the challenge:

> I wish that poets were able to give straight answers to straight questions, but that is like asking astrophysicists to make their calculations on their fingers, where we can watch the process. What I would like to know is: how should I feel about the earth, these days? Where has all the old nature gone? What became of the wild, writhing, unapproachable mass of the life of the world, and what happened to our panicky excitement about it?

And if science today needs anything, it needs to move out of its insular objectivity, its pretense that it deals only with facts, not with ethical implications or free-market motives. What science creates is not only fact but metaphysics—it tells us what we believe about the nature of our existence, and it fosters ever new relationships with the unknown, thereby stirring the deepest waters of our subjectivity. The critics of science are wrong in saying that because of its requirements for objectivity, rigor, and analysis science has robbed us of wonder and reverence. The methods may at times be deadening, the implications spiritually and morally unsettling, the technology frightening, but nowhere can one find more sources of renewal than in the marvels of the material world, be they stellar or cellular. As Karl Popper put it, "materialism has transcended itself" in unveiling mystery after mystery of process and velocity and transformation in even the dumbest rock.

The problem is the speed at which scientific knowledge is growing and the widening distance between those who have a grasp of that expansion and those who have not a clue as to its significance. During the past three hundred years, E. O. Wilson and Charles Lumsden point out, science has undergone exponential growth, meaning the larger its size, the faster it grows. In 1665 there was one scientific journal, the *Philosophical Transactions* of the Royal Society of London; now there are one hundred thousand. In the seventeenth century there were a handful of scientists in the world; now there are three hundred thousand in the United States alone, and scientific knowledge doubles every ten years.

J. Robert Oppenheimer—theoretical physicist, head of the Manhattan Project that developed the first atomic bomb, opponent to the nation's postwar nuclear policy—was a man who had good cause to contemplate the ethical implications of scientific advance. In 1959 he delivered an eloquent talk titled "Tradition and Discovery" to the annual meeting of the American Council of Learned Societies, in which he spoke of

> the imbalance between what is known to us as a community, what is common knowledge, what we take for granted with each other, and in each other, what is known by man; and on the other hand, all the rest, that is known only by small special groups, by the specialized communities, people who are interested and dedicated, who are involved in the work of increasing human knowledge and human understanding but are not able to put it into the common knowledge of man, not able to make it something of which we and our neighbors can be sure that we have been through together, not able to make of it something which,

> rich and beautiful, is the very basis of civilized life. . . . That is why the core of our cognitive life has this sense of emptiness. It is because we learn of learning as we learn of something remote, not concerning us, going on on a distant frontier; and things that are left to our common life are untouched, unstrengthened and unilluminated by this enormous wonder about the world which is everywhere about us, which could flood us with light, yet which is only faintly, and I think rather sentimentally perceived.

Another point of contact: sentimentality is the enemy of both science and poetry.

I have in recent years been interested in the idea of the sequence, both as a poetic form, and metaphorically, as the word is used to describe both the life cycle of a star and the arrangement of genes within the chromosomes. The poetic sequence, as a contemporary form, aims for a kind of fragmented connectedness in a long series of poems or a combination of poetic lines and prose; perhaps it exemplifies the idea that within chaos there is an inherent propensity for order. My book, *The Monarchs: A Poem Sequence,* was inspired by the migration behavior of monarch butterflies and is an extended meditation on intelligence in nature and the often troubled relationship our species has with itself and others. This excerpt will stand as my evidence that careful examination of fact yields easily to contemplation of the miraculous, that a mode of questioning we associate with science can become a nest for poetic delight:

> A caterpillar spits out a sac of silk
> where it lies entombed while its genes
> switch on and off like lights
> on a pinball machine. If every cell
> contains the entire sequence

constituting what or who the creature is,
how does a certain clump of cells
know to line up side by side
and turn into wings, then shut off
while another clump blinks on
spilling pigment into the creature's
emerald green blood, waves of color
flowing into wingscales—black, orange,
white—each zone receptive only to the color
it's destined to become. And then
the wings unfold, still wet from their making,
and for a dangerous moment hold steady
while they stiffen and dry, the double-
layered wing a proto-language—one side
warning enemies, the other luring mates.
And then the pattern-making cells go dormant,
and the butterfly has mastered flight.

In ecology the term *edge effect* refers to a place where a habitat is changing—where a marsh turns into a pond or a forest turns into a field. These places tend to be rich in life forms and survival strategies. We are animals that create mental habitats, such as poetry and science, national and ethnic identity. Each of us lives in several places other than our geographic locale, several life communities, at once. Each of us feels both the abrasion and the enticement of the edges where we meet other habitats and see ourselves in counterpoint to what we have failed to see. What I am calling for is an ecology of culture in which we look for and foster our relatedness across disciplinary lines without forgetting our differences. Maybe if more of us could find ways to practice this kind of ecology we would feel a little less fragmented, a little less harried and uncertain about the efficacy of our respective trades and a little more whole. And poets are, or at least wish they could be, as Robert Kelly has written, "the last scientists of the Whole."

Mastery for human beings is no mere matter of being the animals that we are; we will always push the limits of what we are because it is our nature to do so. The human soul is an aspect of being that comprehends no boundary, no edge. And while the world's nature will always remain evanescent to us, no matter what we do to pin it to the page, we will always find new instruments, such as electron microscopes and literature, with which to gauge the invisible.

*Alison Hawthorne Deming works as director of the University of Arizona Poetry Center. Her previous books include* Science and Other Poems *(Louisiana State University Press, 1994), which won the Walt Whitman Award of the Academy of American Poets,* The Monarchs: A Poem Sequence *(Louisiana State University Press, 1997) and* Temporary Homelands: Essays on Nature, Spirit, and Place *(Picador USA, 1996). This essay is excerpted from a forthcoming book,* The Edges of the Civilized World, *to be published by Picador USA in fall 1998.*

*This essay was selected as the winner of the Bayer Creative Nonfiction Science Writing Award.*

# Astronomy 111

## Grief and Memory

CAROL SANFORD

In my lap is my son's blue spiral lecture notebook, the only one he decorated. At the top of the first page, he wrote ASTRONOMY (with a slanted *A*) in bright orange letters and shaded in behind with an azure-colored pencil. He earned an A in the course in 1983. Five years later he became an air force pilot. His name was Kirk.

***1/17. Cosmic Evolution. The Big Bang was 15 billion years ago! The Oscillating Big Bang Theory: Eventually, gravity halts expansion of the universe, then we get a re-collapse to another Big Bang . . .***

"Come out to see the stars," Glenn calls softly. He's standing outside the door of our little weekend cottage, looking up into the night. "They're so beautiful," he encourages. But I know I'll glance upward, feign appreciation and end up weeping against his shoulder. So I stay on the sofa and continue to read.

***1/28. Earth: 5th largest of the sun's 9 planets. Nearest star is Alpha Centauri. Distance 4.4 light years (1 light year = 6 trillion mi.). Our own galaxy—part of a small cluster of galaxies (about 22) called the Local Group. Sister galaxy (closest to Milky Way) is Andromeda. MW***

***and A each contain billions of stars. (Stars on a clear night—"naked eye giants"—are far away!)***

When we were children, my sister, younger brothers, and I lay out on the grass summer evenings to study the stars. We wrapped up separately in old blankets, looking like Egyptian mummies, because the mosquitoes were mad for blood. With only our eyes exposed, and lying close in a row but unable to see each other unless we rolled from side to side, we stared straight into the stars. We speculated on their number—millions, billions, trillions, quadrillions—and then experienced the giddy effect of realizing no one can ever grasp how impossibly big it all is. Yet here we were lying on this ball called Earth that hung in space like a Christmas tree ornament—attached to what? These thoughts drove us to irreverent acts, like singing "Ninety-nine Bottles of Beer on the Wall." Such counting calmed us.

Our parents gave no instructions about God, and no one in the family knew the constellations except for the Big and Little Dippers, but there was the Milky Way to wonder about and the moon, of course. We tried to find the nose of the man in the moon; we sang "I see the moon, the moon sees me. . . ." One night when Dad was with us on the lawn, he said there would be a man on the moon in our lifetime. I wondered how anyone could actually fly to the moon and walk around up there. It was 1950.

***2/16. Moon. We've got 7 or 8 pounds of moon rock. The moon's motion is eastward among the stars—covers up stars as it moves east (occultation). Dark areas are lowlands (oceans of solidified lava—basalt). Light areas are high lands. Note: The moon was never part of the earth.***

*Luna, luna.* I remember the moon at Mesa Verde. In 1976, Paul, my husband then, and I took our children—Kevin, Kirk, and Renee—to see the Anasazi ruins. A place alive with the ancient and holy. That first night in the campground I saw a sky pregnant with stars—big, brilliant stars that seemed only a story or two above my face as I lay breathlessly watching from my sleeping bag. I could hear silence among the heavenly bodies, and for a moment the bone-white full moon made love to me. An intimate spiritual act. The hundreds of egg cells in my ovaries—magically placed there when I was conceived, and still quietly waiting—mirrored a small cluster of stars in the heavens. I felt connected to everything.

***2/21. Celestial Womb—Birthplace of the stars—middle belt of Orion. Note: All celestial objects move parallel to the celestial sphere.***

When my friend Pat died of cancer, I thought I learned most of what there is to know about grief. But it was nothing like this—my child suddenly dead. For Pat, my mourning began before she died; I wrote poems that said good-bye:

> I lie in an open casket
> unhampered by the dark cake of earth
> my gaze fixed on Orion's belt—
> three well-placed eyes of someone good.
> I know I'm in the heart . . .
> I smile far down inside myself,
> a child content in the womb . . .

I went to the hospital to see her the evening she died. The door to her room was slightly open and she sat up straight on the bed, her back to the door, talking animatedly to her husband. Her head was perfectly bald from radiation and white as the moon. She had never let me see her without her wig; I didn't go in. But that night, I kept her two young children at my house, telling about her love for them until they fell asleep in my arms. Can she now find my son? Give him my love?

*2/28. 3 close stars (1) Vega (2) Altair (3) Deneb. Most of what we see are hot young blue stars. Blue stars are more energetic than red (at every wavelength). Gas/dust causes a star to appear red. White dwarf stars are (1) very abundant (2) very faint (3) very small (size of earth).*

Dwarf, elf, imp, troll. When Kirk was placed in my arms brand-new I was surprised to find he wasn't beautiful in the way his older brother had been. His body was wiry rather than round, and his face was not cherubic but looked like a little old man's: an old soul. Soon his head, with a large occipital lobe at the back and a forehead that protruded over his eyes, seemed misproportioned. He was not really cute, but years later when I saw the movie E.T., there was Kirk at four months: the shape of the head and those round, deep-set eyes—adorable in his own way. And as Kirk grew he was, like E.T., bright, mischievous, and sweet, all at once. I recall a classic scene in the movie: E.T. looks longingly up at the star-filled night sky and yearns to go home—to somewhere certain there in the cosmos, a place he remembers well.

When I rocked my infants to sleep, I had two tunes in my repertoire: "Lullaby and Goodnight," which I mostly hummed because I didn't know all the lyrics, and "Twinkle, Twinkle Little Star." Sometimes I substituted a word: "How I wonder *who* you are / up above the world so high / like a diamond in the sky." A smooth, round forehead pressed to my cheek as we rocked—that is the sweetest memory I have of those years. Together, we made a constellation—a deep glowing red star for my cheek, a glittering blue star for my baby's forehead, and a small white star for a rung of the rocker. What name should we have, forever there in the sky?

*3/2. The Sun (109 times bigger than earth/diameter) is a G2 star—5 billion years old—5 billion to go. Solar wind is the cause of (1) northern lights (2) comet tails.*

All through Kirk's childhood, we camped for a week or two in the summer on Lake Charlevoix. The children swam joyfully all day in the clear blue-green water, and the sun gave us perfect mornings, perfect afternoons. One July night, sitting around the campfire, we suddenly realized the sky was pulsating with red and green, as if a strobe light had been switched on. Every star had surrendered the floor to the flashier, one-night entertainer with that sensational name: Aurora Borealis.

During those vacations, we saw falling stars almost every evening. "There goes one," somebody would say. But it always seemed like the death of something to me—Icarus maybe. And now those lone flights are Kirk, as if he were trying to get back to earth.

***3/14. Pythagoras, Aristarchus, Hipparchus, Ptolemy, (Dark Ages), Copernicus, Brahe, Kepler, Galileo.***

And Van Gogh. I often think of his painting *Starry Night,* and those words in his letter to his brother: "That does not keep me from having a terrible need of—shall I say the word—religion. Then I go out at night to paint the stars." And I remember those lines in Anne Sexton's poem about the same painting:

> . . . This is how
> I want to die:
> into that rushing beast of the night,
> sucked up by that great dragon, to split
> from my life with no flag,
> no belly,
> no cry.

I think my son's death was like that. His jet exploded in a fireball that looked like a Van Gogh star: yellow and green—alive—and rolling, rolling.

***3/23. What is the overall structure of the universe? 3 layers: (1) Solar System (2) Galaxies—all are expanding away from each other! They can go through each other with no star collision! (3) Galactic Groups.***

Kirk wanted to be an astronaut. That was his real boyhood dream. He loved the movie *Star Wars;* Luke Skywalker was his counterpart. Like Luke, he believed in the Force—in the spiritual nature of the universe and in the necessity of fighting evil: Darth Vadar, the dark side of the self, and the Death Star, the outer enemy. He became a fighter pilot after college and proved himself an excellent aviator in competition with his comrades. He loved the men he flew with and they loved him. It seemed there was no stopping him, no end to what he might do.

Less than a week before he died, he called and said, "Well, Mom, I sent for my college transcript today." He didn't have an engineering degree and would have to get one if he really were to attempt to become an astronaut. His request must have been received by the registrar and the transcript sent out to him in Germany, undeliverable. Where would it be? Perhaps it blew out of a postman's hand or wind swept it from a mail truck parked on an airport tarmac. Or it has disguised itself as those envelopes from American Airlines that keep coming to our house, addressed to Kirk, marked "Get a Free Companion Ticket"—third-class mail that I neither open nor can bear to discard.

***4/1. Planets—"close" to stars compared to star distances from each other. Spring Evening planets: Western sky: Venus, Mercury (horizon), Mars. Eastern sky: Jupiter, Saturn, Uranus, Neptune, Pluto.***

Kirk was born a Gemini—under the constellation of the twins Castor and Pollux, sons of Leda. She was the sister of Clytemnestra, whose son, Orestes, murdered her and went mad. The twins were constant companions, like my boys were. When Castor was killed, Pollux—an immortal fathered by Zeus—begged to die too. The brothers were placed in the sky side by side, and they are best seen on April 19, Glenn's and my wedding anniversary. (How well I sense this: the quickest route to madness is to give in to the fear that I might lose my remaining son or daughter!)

Kirk died on April 22. The Sky Calendar tucked in the blue spiral notebook says this night and the next day are marked for celebration: Astronomy Day. And it shows me that the heavenly bodies to look for on April 22 are Pleiades, Mercury, Betelgeuse, Hyades, and Venus.

Kirk was buried on May 5, the day he was to get the new ranking of captain. "It's no big deal," he'd said to me on the phone. Two days after the accident we learned he would receive the promotion posthumously, and his old school friends began to smile and refer to him as Captain Kirk. They had all grown up on reruns of *Star Trek* and were pleased that the military plaque on his gravestone would read Captain.

***4/6. Venus. Sister planet. Planet of love. "Think of Venus, think of clouds."—3rd brightest object in the sky. Can cast a shadow. Comes closest to earth (closer than Mars). Is extremely hot (Hell). Why is Venus such a terrible place? (Close to the sun.)***

In college, Kirk met Jamie—the sister of his soul. Her smiling green eyes appeared in his dreams so he asked her out, and love came easily. Like Venus and Earth's, their relationship was close. But some irreconcilable difference kept them from marrying—career incompatibility or maybe something more basic; I never fully understood. There was hope because they agreed to remain friends. But love became a difficult place for Kirk; wherever he was, he wrote and telephoned her—even after she announced her engagement to another. She married, then he too found someone and married—someone lovely—and soon after, he was dead. Now, though, when Jamie flies she has no fear of dying.

***Note: Apr. 25, 7–8 p.m., Brooks 176. SPECIAL LECTURE ON BLACK HOLES.***
***4/13. Law of Universal Gravitation—works over infinite distances. Gravity is caused by mass. ("In orbit" means the velocity is just right so that you fall around the planet.)***

The air force eventually concluded that Kirk's accident was due to "gravitational loss of consciousness," in other words, blackout; and I will always wonder if he was conscious in the last seconds. I say the Earth called him back to itself. It was Earth Day. Onlookers in the little German town saw the jet bank to the right then go down gracefully, swiftly. Someone thought he could see Kirk in the cockpit. When the jet hit in the woods, a shallow crater was blown in the earth and fire destroyed the trees. Rain began the next day and continued for six more. The water that filled the crater was polluted by fuel.

After gathering what was left of Kirk, and after a week of collecting pieces of metal, the clean-up crew held a memorial service of prayers, scripture, and hymns. The local "forestmeister," with townspeople standing around him, presented a cross he'd carved from a nice piece of wood. It was placed on a knoll at the edge of the crater along with a special marker made to look like the front of a little Bavarian cottage:

> *Zum Gedenken*
> First Lieutenant
> Kirk P. Shauger
> 22, April 1992

*Zum Gedenken:* In our thoughts. The next day the townspeople planted trees to eventually cover the earth's scar. Life went on, as it must.

*Missed April 22 lecture. (Mars)*

My throat goes dry when I take this in: He missed class in 1983 on the same date of his crash in 1992. Then too, the planet Mars, the god of war. . . .

*4/25. Black Holes—Huge! Spin very fast so light cannot escape. (Gravity is so strong that the escape velocity equals the speed of light.) Kinds—a) mini b) stellar c) galactic centers. There are 2 confirmed Black Hole stars.*

Yes, it's true. A child's death equals black hole: Instantly nothing matters—just as matter is nothing to the black hole. It swallows up hope, awe, interest, desire, appetite, pleasure, joy, goodness, sometimes love itself. There's no escape. No light. You are face to face with the absolute.

I read this somewhere: Is it possible to transform yourself after you have suffered the greatest loss you could ever imagine? I want to know, do time or grace give back enough to live by?

*4/27. Possible test questions. Where are we? Is there life in the atmosphere? What is the overall structure of the universe? Why do stars get old and eventually die? What are spiraling arms (really)?*

I'll take the test. Kirk, I'm here and you're there; we are separated. Even in the mass of air surrounding the earth, your spirit and others move at will and see us, the living. While you are fully enlightened, from here it seems the overall structure of the universe is dark mystery with only occasional light to see by. I myself am an aging star who will eventually die, worn out by the losses of life. But I will return to the source of all things and find you there. As for spiraling arms, they are really these arms—mine. Human, they long to hold you now. They want to reach out forever, growing thinner and longer, like the pale stardust of asteroids, and sweeping wide, draw you into myself: my child.

I walk out into the yard of our little cottage and take Glenn's hand. Remembering some words from early childhood, I say them as an innocent—out loud, invoking their power:

> Star bright, star light
> First star I see tonight
> I wish I may, wish I might
> Have the wish I wish tonight.

I want to live again.

*CAROL SANFORD, recently retired from university teaching, is currently at work on a memoir about building a log cabin with her husband in central Michigan. Her poetry has appeared in* Plain Songs, Verve, *and* Parting Gifts.

# Earthquake Country

SUSAN MANN

Most of us live hidden lives. Who shares a romantic breakup with a coworker, or tells their child fears of an impending layoff? Those of us with the voices of schizophrenia are doubly hidden: in addition to the normal masks of work and family life, we must never betray that we hear threatening voices as clear and vivid as a warning called out by a friend.

For many years, I was a successful lawyer. Most days, I surfed the Net with the best of them, faced off my opponents with delight. Other days I was inexplicably absent, or called in sick for weeks with a "stomach virus." I wanted to say, "The medication killed my white blood cells, and that has damn well near killed me." I wanted to say, "The voices are nonstop today." But of course I did not speak. I had clients. I had a license.

I grew up in earthquake country. The quakes I experienced there have always been my metaphor for mental illness: life as a series of dislocations. The year I was seventeen, I occasionally "heard" my older brother plotting to kill me.

I first realized I must be seriously mentally ill when in that same year, alone in my employer's house, chance had caught me close to the epicenter—the place where the plates are actually moving and so of greatest force—of a significant earthquake. The type of quakes routine along the San Andreas Fault feel very much like the moment in surfing where the wave takes you, both in immediacy and force, and for that reason I had always enjoyed them. But I had been drilled in school to respond instinctively to movement of just this violence. I knew I had to get to a threshold (because the ceiling might come down on me) and then, in the breathing moment when the movement stopped and I wanted to hug safety, to get outside very quickly. I got to the doorway. But I froze. Nothing had prepared me for the sight of the ground moving toward me, wave-high, as the aftershock hit.

By the time I was twenty-three, my first year of law school, the threatening voices had become constant. I heard my classmates plotting against me whenever I left my dorm room for class or the library. The threats would often drown out the real voices of gossip and instruction. Many of the "voices" I heard were those of my friends and study partners. They had no reason to hate me.

One can leave the faultline. I moved East. But one cannot leave the country of psychosis. For schizophrenics, the moments where the earth lurches violently away just continue. We can only learn to dive for a threshold and make some joke to cover the movement. Those who care about us teach us such camouflage; I was about to embark on a course in the saving powers of denial.

It was twelve years before someone gave me a name for the voices. I did not seek treatment in those first five or six years. I spoke to no one of the tremors that repeatedly jarred the ground below my feet. Had someone pressed me about this choice, I would have said that, at that time, a diagnosis of active schizophrenia would have precluded my admission to the bar. How could I have known the wisdom of silence at seventeen? My family had a secret: my older brother.

By my mid-teens, my brother, ten years my senior, had slipped into illness and returned home after several years on his own. My parents mourned his illness with grief, love—and ultimately silence. To this day, they will not name my brother's diagnosis. We did not discuss his symptoms—much less my growing fear that I shared them. My parents simply dropped their active social lives when he returned home.

The closest they came to acknowledging his condition was my father's anger at the dinner table one evening. Our town paper had run a profile on a local college student's struggles with manic-depression. "Why did that fool let them exploit him like that?" he asked. "He'll never get a clearance now. And if he goes on to the law school, he's killed his chances of admission to the bar." In my father's view at that time, as he put it to me when I first told him of my illness at twenty-eight, overcoming mental illness was like surmounting alcoholism: a matter of will to be conducted with decent reticence.

I believe that my parents did the best they could with a situation compounded by love and anguish. Certainly, after the initial shock, they have been supportive beyond any expectation with respect to my own illness. But Charles's crisis struck him twenty years ago, before the spate of books and movies on mental illness that have made it an acceptable topic of conversation. And the intimate knowledge of the stigma that my family gave me was the best tool they could have offered me for a successful adult life. Fearing the silence of family and friends, I hid my illness from everyone but my doctor until I was twenty-eight and from most others until the progression of my illness outed me. Because very few people knew, no one discouraged me or excluded me from law. This gift—of knowing when to hide—was the most important that I ever received, for it made possible my practice of law. And it was the skills and money that I gained in practice that provided the kind of intensive treatment that allowed me to keep working.

The voices eventually drove me to seek treatment at twenty-four. The constant soundtrack of threats of physical violence and of professional ruin had begun to affect my professional life. I often heard my colleagues discussing their intent to have me fired for offenses I had not committed, such as falsifying my time sheets. Less commonly, I also heard unknown voices screaming that I was stupid, ugly, and useless. The fear these soundtracks inspired was not the main problem. Many people with schizophrenia learn to live with such fears. I had even learned to speak effectively when the voices were so loud that I could not hear my own words, provided I didn't have to take questions. At that time, I could also identify the falsity of a hallucination 80 percent of the time if I had time to test it against what I did know. In law school, I had coped with the other 20 percent by keeping my mouth shut.

My first job as a summer clerk in the law department of a major corporation showed me that evading treatment now cost more than it gained me. The pace of litigation does not lend itself to introspection, especially if you have no idea what you are doing; no client or supervisor can allow you to speak purely at your option.

One of the department's experienced lawyers asked me to participate in taking a deposition. In this deposition, as often happens, the witness' lawyer tried to deflect attention from his client's testimony by shouting challenges to many areas of questioning. Since the stress of hostility exacerbates my voices, I felt like someone trying to join in a conversation while listening to loud rock on headphones. I could not follow the arguments over what questions the witness would be permitted to answer. Confused and overwhelmed by the clamor of the opposing counsel's heckling and the angry shouts of my auditory hallucinations, I passed when it came my turn to speak. This situation was multiplied many times by the end of the summer. My supervisor took me to lunch to tell me, "Your written work was excellent, but we just can't hire a lawyer so handicapped by shyness."

Having had that experience, when I first walked into psychiatrists' offices, I put the problem as sharply as I could. I told the doctors I suffered from auditory hallucinations that consistently interfered with my work performance and that I had a brother on federal disability for mental illness. I walked out with the offer of psychoanalysis, cognitive therapy, or support. These talking cures were so different from the relief from the voices that I was seeking that I didn't actually end up in treatment for three years more and went without medication for a year beyond that.

But I have come to see that it was probably kindness rather than perversity that led those men to offer me talk rather than pills, that led to diagnoses like "social anxiety." Or maybe not even kindness, maybe the kind of instinct that leads one to jerk a child back from the edge of a cliff. Within a few years of medication treatment, I came to suffer the visible side effects of antipsychotic medication from time to time, among them, tardive dyskinesia (TD), a form of involuntary mouth and tongue movements chiefly caused by antipsychotics. I also suffered from myoclonus, a form of muscle jerking that often led me to drop things or, occasionally, to fall down in public, on the subway, or in my supervisor's office in the middle of discussing a brief. I cannot always hide that I am on antipsychotic medication. In the early days of my career, medication might have cost me housing, friendships, and the discipline and possibilities of adult work. It would have meant what Susanna Kaysen evoked when she titled her book about schizophrenia *Girl, Interrupted:* disruption at the very point at which an adult self begins. I did not understand the penalties that could follow medication. The doctors knew.

Of course, in seeking treatment for mental illness at all, I was rejecting what I already knew of the penalties of accepting a diagnosis. Despite the difference in our ages, Charles and I had been close while growing up. He had introduced me to all the mysteries of our childhood: the hawks in the foothills and the rock of the high Sierras, every cove along our coastline and the bookstores of the city. Before the onset of his illness in his late twenties, he seemed a mature adult in every way to me. He worked full-time. He had a book accepted for publication. He could drive.

For twenty years now, Charles has been on SSI, the federal disability program. He chooses not to be on medication. My best guess is he would be well-employed if he had the kind of access to medication and advice money buys. Some years ago, he went to one of the good music programs and composed lovely, lambent music. For now, he lives on the five hundred dollars a month that SSI gives him. He lives in a rooming house in the grim center of Los Angeles. He can't afford to ride the public buses. His is the story I have been fighting for twenty years not to make mine. I grieve for him. Not wealthy myself, I can offer him no route off the faultline.

That faultline now rules my life. It was four years ago that I first had any sustained experience of a psychiatric hospital, when I participated in a research study that required a stay of several weeks. I had expected to experience the openness of an intensive therapeutic relationship as a relief after my double life. But I discovered that I had, if anything, underrated the virtues of silence.

Being with people who knew my diagnosis was far more constricting than suppressing my experience as I did outside. One nurse took me to task for not making contingency plans for losing my job. A social worker sat me down with a newspaper and commented, "You know, an efficiency to yourself downtown really is a luxury. You could pay half that for a room further out." She painstakingly "showed" me how to read the Rooms-for-Rent section. Another social worker suggested I move across state lines because the public benefits would be better. All these comments were well meant—and responsive to my (then unfounded) anxiety about my job—but they undercut my sense of myself as competent just when I most doubted it. Since many schizophrenics are on the federal programs for the disabled, a major role of a psychiatric hospital is to confront patients with the true reality of their situations, to encourage them to seek long-term housing and treatment they can afford. But it often works best to honor the possibility in people.

When I first worked as a lawyer, I would offer a legal solution to every problem someone put before me. But I came to realize that this was both unhelpful and unwise. For example, an entrepreneur might come into my office saying he needed a licensing agreement when what he really wanted was to talk to a lawyer about his business. Speaking of the licensing agreement gave him a neutral reason for asking around for a referral. But in fact he might simply want to be incorporated and know how quickly I could "paper" an introductory public offering if he got venture capital. And I learned not to express an opinion on whether one business venture or another would yield the best return to capital; the very thing that makes for a successful entrepreneur is the ability to push the envelope of the possible. In a sense all I'm arguing for on the behalf of people with schizophrenia is that moment of listening—that moment when one person speaks and the hearer listens for the other person's own sense of the necessary and the possible.

That sense of possibility is one important thing I lose when people learn I am a schizophrenic. The question people most often ask me is, "Can you live alone?" Or even, "Are you violent?" I want to respond, "Of course, I'm an adult," or "Of course not. Are you?" But I forfeit adult status whenever my illness becomes known.

More than most stereotypes, those behind these questions combine fantasy and fact. Many schizophrenics do live with relatives or in group homes. While auditory hallucinations are the best-known symptom of schizophrenia, the cognitive disorganization also characteristic of the illness makes it difficult for many to manage money and handle the tasks of daily life. In my own worst month, I overpaid or underpaid all my bills, sending Bell Atlantic $330 for a $23 charge, and found myself unable to navigate the city, misplacing the grocery store where I had shopped for eight years. Moreover, as is well-known, many of the mentally ill are homeless, in part because it is so hard to gain a place on the federal disability rolls.

In a small percentage of patients, paranoid delusions can lead to violence. When the voices announce someone is about to murder you, for example, one response is a preemptive strike. My stays in the hospital included witnessing two such paranoid attacks, and once, in another context, I was on the receiving end of one myself. However, that incident was my lone moment of danger in many years of associating with schizophrenics. A significant number of us meet the normal milestones of adulthood: schooling, an apartment, satisfying work, relations with others based more on friendship than on fear. Because we are so little known, our only and best defense is to conceal our illness.

The key to concealing—and living with—schizophrenia is effective medication management. Although I have been on medication for eight years, it has become less and less effective for me over the past four. In this I fall into the minority of the schizophrenic population who are "drug-resistant," that is, whose symptoms do not respond well to medication treatment even when this is zealously pursued.

At first, medication served me well. Having stumbled into an antifraud practice, I was happy in my work and lucky in my friendships. But, even in the early years, side effects introduced me to the ambiguities of choosing appropriate antipsychotic medication. After only a year on Mellaril, one of the traditional antipsychotics, I developed TD and chose to terminate the medication. I made this choice because tardive dyskinesia, unlike most other side effects of antipsychotics, often does not go away when the medication is withdrawn. The risk of permanency increases the longer one stays on the medication.

As I would later come to experience it from time to time, TD's involuntary tongue movements made eating difficult and left me in constant pain from the sores I had chewed in my tongue. When the tongue movements moved outside my mouth, even my mother found the bizarre, stereotypical movements stressful. These are not problems one would wish to embrace for a lifetime. First as a high school and college athlete, then as a runner and cyclist, I had always been at home in my body. To publicly and painfully lose control over my own face, tongue, and jaw shook the ground beneath my feet. It reminded me of the time I walked around my hometown in the aftermath of the

Loma Prieta quake (the one that horrifically collapsed the freeway). One building I knew well had been reduced to cement debris. The earth should not move.

Like most people who develop TD and have generous private insurance, I simply switched to Clozaril. Clozaril, the only approved antipsychotic that does not cause TD, is an expensive drug, both in itself and in the monitoring required. Because it can cause a reduction in white blood cells (WBCs) that is fatal if left unchecked, its manufacturer requires its users to have weekly blood tests and to provide a weekly psychiatrist's signature attesting that the patient maintains a safe white-cell count. The cost of Clozaril and the monitoring amounts to seven or eight thousand dollars a year.

Initially I found Clozaril's side effects—muscle stiffness and jerking, weight gain, somnolence requiring twelve hours of sleep a night—a fair trade for reduced psychotic symptoms. Even though the muscle jerking led to my once bouncing down a subway escalator, seriously injuring my back, and to my hurling every breakable dish I owned to shatter on my kitchen tiles, I chose to persist in treatment. The peace in my work life and in my avocation of writing repaid the stress of treatment. By this time, I had published ten articles in magazines and newspapers, drafted winning briefs, received an offer of marriage. I shared my sister's joy in her young daughter and cared for her often. I had had a taste, however tentative, of adulthood, and I was not about to give it up.

However, four years ago, I was unfortunate enough to be in the minority of patients who develop the white blood cell problem. My WBC count dropped below safe levels and stayed there for six weeks; I was abruptly removed from the medication. I lost thirty-five pounds in four weeks and have never been so physically ill in my life. I felt an exhaustion I had only felt before in the lassitude of pneumonia, and even simple activities, like turning on my bedroom light or standing up, led me to vomit convulsively.

Having tried the other FDA-approved drugs without success at that time, my doctor referred me to the research trials of investigational antipsychotics available in a nearby city. In the past several years, I have tried most of the experimental drugs being tested as well as the new antipsychotic the FDA approved last fall. I am now willing to experience significant side effects, including some tardive dyskinesia.

I also have been willing to spend the extended time in a psychiatric hospital the three research studies required. Whenever I hear the word *hospital* these days, I smell air breathed too many times and hear the sounds of a locked ward: nurses shout orders, doors clang, cries ring out. A patient bursts into my room repeatedly one night after midnight, convinced I have hidden her daughter. Another night I wake to find a male patient in our room in the process of lifting my roommate's jacket; soon thereafter I walk into our room to find him curled up on my bed. One patient in a room near mine beats his head against the wall and rips the sink from his bathroom wall. Later he tells me he will break another patient's neck, pounding his fist into his palm for emphasis. At the twice-daily community meeting, we must state whether we are suicidal or homicidal. The answer is not always "no." Since I am, I am invited to visit the Quiet Room to shout and cry over my situation, because it would allow me to experience my "essential self." But, for me, having shaped an adult self through law practice, community work, and the regular care of a child, Carolyn Heilbrun expressed it best when she wrote, "The essential self lives in the work that commands the full attention." Attention directed outwards, not toward the self, and toward the appearance of normality, has meant sanity for me. With the constant talk of symptoms in the hospital, this outward focus can be lost and with it much ability to function.

Finally, although I have followed my doctors' every suggestion, and tried others on my own—yoga, meditation, acupuncture—I can no longer work. The hallucinations and cognitive confusion are so constant they give me two to three useful hours some days—other days not so much.

Since I currently have a lawyer's savings, disability insurance, and good health coverage, my inability to work is a misfortune, not a tragedy. For others, however, the tragedy is real.

I was simply lucky. Schizophrenia strikes many people in their late teens and early twenties, before they have had a chance to acquire job skills. Even for those who gain them, the Americans with Disabilities Act (ADA) is sometimes of little help. The ADA addresses the needs of those who can work with a reasonable accommodation—say time off weekly to see a psychiatrist—rather than those who are, like myself and many schizophrenics, unable to work at all during extended periodic crises. And to reveal that you suffer from mental illness, as seeking accommodation under the ADA requires, can itself jeopardize a job, though legally it should not. Finally, some never respond as well to medication as I had in the first years of my treatment. Those without a substantial work history for any of these reasons are left to the basic $484 a month SSI provides.

The comparatively slow onset of my symptoms was also important in determining my treatment options. When I first entered treatment, I was already practicing and could afford any medication, any level of intervention. I often was able to see the psychiatrist who wrote my prescriptions weekly, not just for the brief visits every one to two months some people on Medicaid or in some HMOs must rely on. This allowed us to fine-tune medication options quickly for the best functioning. I could also try any drug on the market whenever my doctor advised it, even those that cost several thousand dollars a year. Not every patient in an HMO or on Medicaid has these options.

Certainly, learning to advocate for my clients' interests taught me how to protect my own. Antipsychotics and the drugs that control their side effects can affect memory and thought, particularly at the higher doses. When I first met the psychiatrist who treated me for several years, I was on 50 mg of Clozaril, far below the 300 mg seen as the threshold effective dose. My doctor only consented to accept me as a patient on the agreement that I would increase the medication to 300 mg. I tried 300 mg and experienced a dramatic reduction in auditory hallucinations, but this dosage left me too sedated to think well enough to do my job. It took every negotiating skill I had learned in law practice to convince my doctor to accept that I would remain, long-term, on 225 mg, a significantly lower dose that allowed some voices to break through. On this and other medication issues, I recited, "my mind is what I sell," in his office and those of other doctors. That I made my living as a professional was what gave this mantra its plausibility.

It is hard for those off the faultline to understand how routine the quakes are for those of us who live on it. The apartment buildings that will not rent to me. The stares on the bus and rejection from employers for those who suffer from TD. Friendships and love relationships lost as soon as the illness becomes known. If only I could simply state my illness just as I announce I come from California: one attribute among many, not a total definition of self.

***Susan Mann is a writer and lawyer living in Washington, D.C. Her ten previously published articles have appeared in the* New York Times, *the* San Francisco Chronicle, *and the* Stanford Law Review, *among others.***

# The Moon on Ice

SCOTT A. SANDFORD

On January 13, 1989, I was standing in the wind on a blue ice field near the MacAlpine Hills in Antarctica, about 360 miles from the South Pole. I was there to find and collect meteorites. By this time I was something of an old hand at hunting for meteorites in Antarctica. I had already served as a member of the 1984–1985 austral summer Antarctic Search for Meteorites (ANSMET) expedition during which we had found almost two hundred meteorites in and around the Allan Hills in Victoria Land. One of these meteorites, ALH84001, was destined to create a major fuss in 1996 when it would be suggested that it harbored evidence for the presence of early life on Mars. However, on this particular day these revelations were still in the future. In any event, I had other things on my mind, not the least of which was the wind.

It's hard to spend much time out in the open in Antarctica and not get intensely interested in the wind. On the Antarctic plateau the wind is omnipresent. It constantly demands your attention through its ceaseless collision with any and all barriers that try to contest its free flow. It is easy, even years after leaving the polar plateau, to vividly recall how the wind flutters and snaps the folds of your anorak, or how it bangs the canvas on your tent, or how the snow sounds as it hisses against your pants legs. The winds on the plateau are not the winds that we have grown up with. Normal winds are the products of weather, high and low pressure systems wrestling for dominance. But the winds of the Antarctic plateau are a different beast. They are katabatic winds, winds driven by gravity.

When air gets over Antarctica, it cools. This causes its density to increase, and it sinks until it hits the ground. In most places on the earth, sinking cold air will ultimately be reheated by sun-warmed ground, or by mixing with warmer air, and the air will get its chance to soar again. But in Antarctica the ground has no warmth within itself and the sun has little power, even in the summer when it is up twenty-four hours a day. When cold, sinking air in Antarctica hits the ground, it still wants to sink. It continues to seek a lower place. In its relentless pursuit of down, it flows along the ground, following the contours of the land. Like a river of air, it seeks the sea. It is a wind driven by topography, not weather.

Such a wind is your enemy. It is nearly always there, and it is what makes the Antarctic plateau so cold. The temperatures on the polar plateau along the Transantarctic Mountain Range are typically -5 to -30°C (23 to -22°F) in the summer. This is cold. But not too cold. With proper clothing and after you acclimatize, these temperatures are not so bad, even when you work outside for fourteen hours and then go home to a canvas tent. Indeed, I can remember one day when the temperature actually got up to freezing, and we all began to feel so hot that we stripped down to our long johns. Of course, that was a rare day. One without wind.

Moving air adds the dimension of wind chill. Properly clothed, a windless -15°C day on the plateau can be downright comfortable, but add a twenty-knot wind, and it can quickly become dangerous. Such a wind bites right through your clothing and turns your gloved hands into unresponsive lumps of ice. Exposed skin can quickly become frostnipped and, if you're not careful, frostbitten. On a bad day you can actually hear the faint snapping sound ice crystals make when they spontaneously form in one of your ears.

Really strong katabatic winds result in truly strange but deadly storms. Since the wind is gravity driven, it consists of only a thin layer of air, maybe thirty feet deep, that is flowing downhill. Within this layer the wind chill may be enormous and an exposed person would be in serious danger of freezing. And yet only a short distance overhead the wind velocity may be almost zero. I once met a navy weatherman who claimed that he climbed an open lattice weather tower during such a storm and that at an altitude of forty feet above the surface there was no storm. Having never had such a tower available when I've been in such a storm, I will have to take his word for it. Still, I think his story has the ring of truth. I've been in storms where the wind was blowing so hard that I couldn't see more than twenty feet along the ground in any direction because of the blowing snow, and yet I could see a blue sky directly overhead.

Blowing snow represents its own special kind of hazard. While little new snow falls in Antarctica (it's one of the world's biggest deserts), what does fall doesn't melt. Instead it accumulates, and the portion on the surface is constantly being driven around by the wind. As a result snowdrifts called *sastrugi* are constantly forming and being destroyed on the surface. In practical terms, this means that you have to be very careful about putting things down. Left unattended for even ten minutes, smaller items can be drifted in and buried. In a land without landmarks, such items are easily lost forever. Tools are the most common objects to disappear in this manner, but larger things have been known to go missing as well. I know for a fact that somewhere up on the Antarctic plateau there is an entire case of scotch that has lost its way.

The katabatic winds in the Antarctic interior are such a part of everyday life that they change the very nature of human interaction. Do you want someone to think you are polite? Then sit upwind from them when you stop for lunch and block the wind. Is it rude to thrust your face into someone else's and minutely inspect their every feature? Not here! In Antarctica this is a sign of friendship and concern. How else can you catch the early signs of frostnip on your buddy's nose or ear lobes? Or consider this. Normally it is considered to be somewhat rude not to look a person in the face as you talk to them. Not so on the plateau. In such an environment, it is frequently best to converse by standing side by side while facing downwind so that everybody's faces are protected by their parka hoods. I have participated in conversations where as many as six people stood in a long line oriented perpendicular to the wind, everyone facing downwind and heatedly discussing an issue, all the while without once looking at each other. So relentless is the effect of the wind that this habit can take some time to completely break, even after it no longer serves any useful purpose. I can remember once standing in a park in Christchurch, New Zealand, shortly after returning from six weeks in Antarctica and being puzzled by the peculiar looks my friend and I were getting from all the passersby. It wasn't until later that I realized that we had been having an animated conversation while standing side by side in our shorts and T-shirts and facing, in parallel, away from a gentle summer breeze!

Now to be fair, a katabatic wind is not always necessarily the enemy. Sometimes it can be your friend. For example, since it is the local topography that guides the downward flow of a katabatic wind, its direction at any given location on the plateau is the same nine days out of ten. In a terrain that consists of an unbroken expanse of snow that goes to the horizon in every direction, and where a compass only tells you which way is down, this can be a godsend for navigation. In a land with no landmarks, the wind can give you something to steer by.

And, in a way, it was the wind that was responsible for my being in Antarctica. I had come to the MacAlpine Hills to collect meteorites, and without the wind there would have been no point in trying. Meteorites fall uniformly over the entire surface of the earth. In principle, this would suggest that searches for meteorites should be equally likely to yield results anywhere you look. However, a number of practical issues associated with meteorite hunting make Antarctica the best place to search. The first practical issue is simple: you cannot collect a meteorite if you cannot see it, and in Antarctica there's nothing to get in the way of your view. No trees. No houses. No grass. No nothing. I once spotted a meteorite in Antarctica from almost a half a mile away.

Antarctica's second advantage is almost as simple. Meteorites must not only be seen, but they must be identified as such before they can be collected. Most locations on Earth are covered with rocks. Depending on the types of rock present, there may be significant confusion between the local junk rock and any meteorites present. Meteorite laboratories routinely receive a steady flow of possible meteorites sent to them by members of the public. The vast majority of these turn out to be charcoal briquettes, lumps of coal, slag from smelters, etc. Again, Antarctica stands out. While it may take a trained meteoriticist to spot a meteorite lying on a old basaltic lava flow, any idiot can recognize a dark meteorite sitting on white snow or blue ice. Most meteorites collected in Antarctica are found on blue ice fields. In these locations there is usually little or no terrestrial junk rock on the surface and virtually every sample you find is a meteorite.

A third advantage enjoyed by Antarctica is that meteorites last longer there. When a meteorite falls to Earth, it immediately begins to be degraded by all the same processes acting on terrestrial rocks. Most meteorites that fall on land are never collected. Instead they are plowed under by farmers, paved over by highway departments, destroyed by weathering processes, and so on. Meteorites falling on the ice sheets of Antarctica, however, are quickly buried in snow and ice after which they are saved from extensive weathering, highway departments, and the lot.

Finally, the ever-present wind plays a key role too. A meteorite landing on the snowy surface of Antarctica is quickly buried by additional snow. As snow accumulates, its weight compresses it into ice. The resulting ice and overlying snow are generally not stationary, but instead they flow slowly downhill. Since they are entrained in the ice, the meteorites go along for the ride. Most of this ice ultimately flows to the coast where it breaks up into icebergs that float out to sea and melt. Thus, it is the ultimate fate of most meteorites that fall in Antarctica to take an ocean cruise at the end of which they are unceremoniously dumped into the sea. Fortunately for meteorite hunters, not all the ice formed in Antarctica makes it to the coast. There are some locations where the moving ice collides with some form of geographic barrier, for example, nunataks, or mountains that are partially or fully submerged in the ice. This can sometimes lead to a condition where the ice stagnates and cannot easily flow around the barrier to the left or the right; instead it is thrust upward. In Antarctica, with its non-stop, low humidity winds, the upward moving ice and snow are subjected to rapid ablation. They sublimate away in much the same way that ice in your frost-free freezer disappears. As the ice ablates away, meteorites entrained within it are exposed on the surface. The result is a kind of meteorite conveyor belt that gathers meteorites from the interior of the continent and delivers them, free of charge, to special ablation zones called blue ice fields, where they accumulate. Find such an ablation zone, and you are in meteorite heaven. And it's all thanks to the wind.

By January 13, 1989, I had already spent over six weeks of the 1988–1989 austral summer season as a member of a party of eight searching for meteorites along the Lewis Cliff Ice Tongue. The weather had, by Antarctic standards, been remarkably good all season and we had been able to work long hours almost every day since our arrival (of course, the sun never set while we were there). We had been wildly successful, having already found almost a thousand meteorites. We were also dog-tired, and after weeks living under adverse conditions and in close quarters in tents, occasionally sick to death of each other. Given that we had already guaranteed a successful season and perhaps sensing that the members of the team needed a break from the established routine, the leader of the expedition, Bill Cassidy of the University of Pittsburgh, decided that it would be best to stop our systematic collection a little early and use the remaining week of the trip to scout for meteorites at several nearby blue ice fields. And so it was that on

January 12, 1989, four of us were sent on a traverse to the MacAlpine Hills, about ten miles away as the Antarctic skua gull flies, but closer to twice that by snowmobile because of the need to avoid intervening cliffs and crevasse fields.

Cross-country traverses can be mildly complicated affairs. Powerful snowmobiles are used to tow trains of one to three sledges along in single file. Since everything one needs to survive in Antarctica must be taken with you wherever you go, the sledges are usually heavily laden with tents, food, clothes, fuel for the snowmobiles, scientific equipment, and survival gear. The sledges can be very heavy and can be very challenging to control over the ice and snow. They frequently flip in rough terrain, and it usually takes the combined effort of two or more people to heave them back onto their skids. Because of the possible danger of hidden crevasses, traverses are always made in single file. The theory is simple—if everyone travels in single file, only the first person can be caught by surprise. If the first in line breaks through the snow bridge of a crevasse, those following have plenty of time to stop, reconnoiter, and, if necessary, effect a rescue.

Things quickly began to go sour. Bill had appointed me to be the nominal leader of the four-person party. Unfortunately, this appointment came with responsibility, but not authority. By this time the other three members of the party—Roberta ("Robbie") Score of the Johnson Space Center, Randy Korotev of Washington University in St. Louis, and Monica Grady, now at the Natural History Museum in London—and I were all good friends (Robbie and I had also worked together during the 1984–1985 field season). I was finding that it was remarkably difficult to lead when the day before we had all been followers together. As a result, I was doing a bad job of giving orders, and they were doing an equally bad job of following them. The fact that we were all tired didn't help.

To make things worse, after setting up camp at the MacAlpine Hills and recovering four fifty-five-gallon fuel drums that had been airdropped nearby for us, we established that the local geography cut us off from radio contact with the other half of the party. We could occasionally contact the base at the South Pole, but the conversations consisted mostly of shouted requests at each other to please repeat what the other had just said. After about a half hour of this we got South Pole to understand that we had arrived at our new location safely and gave up trying to pass on any information of a more complex nature.

We awoke on the morning of the thirteenth to a temperature of -18°C and an eighteen-knot wind. Normally we would have considered waiting for some improvement before spending the day searching, but there wasn't much time remaining in the season. We decided to work anyway. The search started off well. We found our first meteorite of the day only a hundred yards from camp, before we even got to the ice field itself. But our luck was not to hold. While I was following Robbie Score down a steep slope to the field, she crossed a large crevasse and her snowmobile punched out a small portion of the snow bridge. I turned to avoid the spot, but on the steep slope my snowmobile lost traction and slid sideways onto the crevasse. Rather than subject the snow bridge to the entire weight of myself and the snowmobile, I jumped across the crevasse while my snowmobile tipped over sideways on the bridge. No permanent harm was done, but I had hit the ice at about fifteen miles per hour and slid quite a ways down the slope.

The wind proved to be even worse on the ice field. At one point, while writing some notes in my field notebook, a particularly strong gust of wind snatched the book from my arms and almost knocked me from my feet. The papers in the notebook scattered. I ran as fast as I could on the ice, but I only managed to recover a small fraction of them. Most of the escaped sheets probably not did stop until they reached the Ross Sea. They were gone for good.

By this time we were all feeling thoroughly tired, cold, and, given the problems we'd had that day, irritable. Nonetheless, we doggedly continued to search up and down the ice field. It was on one such pass, while driving into the wind, that I found myself executing what I refer to as a "high-speed meteorite dismount."

This maneuver requires some explanation. In principle, searching for meteorites on the ice fields of Antarctica is a fairly simple affair. If you're lucky, you're searching an ice field that is uncontaminated by terrestrial rocks. On these fields, everything you see that isn't snow or ice is probably a meteorite. Under these conditions, vast areas can be swept quickly and simply by having everybody drive their snowmobiles in echelon back and forth across the ice field while the people at the ends of the line make sure that new ice is swept during each pass. Unfortunately, many ice fields contain terrestrial junk rock that is also being carried along with the ice. On these fields, finding meteorites can be a more difficult affair since the searcher must scan every rock and make a judgment call as to whether the rock is a meteorite or not.

With practice, you can usually have a good chance of telling whether a rock is a meteorite just by looking at it. Most meteorites have a distinctive black coating called a fusion crust. The fusion crust is the thin layer of material that was briefly melted by friction with the air when the meteorite fell at high speed through Earth's atmosphere. After seeing many hundreds of meteorites on the Antarctic ice, most searchers develop a subconscious ability to spot a meteorite from a moving snowmobile. Many veterans of Antarctic meteorite searches have perfected this instinctive reaction into the "high-speed meteorite dismount" I mentioned earlier. The dismount goes something like this: (1) spot a meteorite within your assigned strip of ice, (2) immediately stand and draw your off-side foot across the seat of the snowmobile while (3) almost simultaneously taking your thumb off the throttle; the resulting deceleration of the snowmobile will throw you forward and off to the side of the snowmobile, (4) hit the ground running, (5) slide and skitter your way to a stop, hopefully near the rock in question. Properly executed, the snowmobile should roll to a stop nearby on its own. After six weeks on the ice, this entire process can become almost entirely automatic and largely subconscious.

This particular dismount was a relatively ugly affair. The rock in question was fairly large by normal meteorite standards, about the size of a clenched fist, and the meteorite alarms in my head had gone off so strongly that I had dismounted before the snowmobile had slowed down even slightly. Flailing my arms and running as fast as I could, I skittered past the cause of my excitement and nearly lost my feet. By the time I fully regained my balance, I had to backtrack.

When I got back to the rock that had set off my instinctive leap I was a bit puzzled. It was clearly the wrong color to be a normal meteorite. This rock was colored a peculiar tint of greenish brown, nothing like the dull matte black typical of most meteorites. I flopped down on the ice on my belly to get a better look. Hmmm, this was interesting! The color wasn't right, but the surface of the rock had the texture and shape of a typical fusion crust. This was either a very unique meteorite, or a very weird junk rock. I looked around the ice nearby. Rats! There were a few other rocks in sight and they clearly *were* junk rock. A dilemma. Do we collect this rock or not? By this time the others had spotted me face down on the ice and had turned their snowmobiles to gather around the rock. I scrambled back to my feet, and we huddled in a group to discuss the matter.

Now by this time we had all seen hundreds of meteorites on the ice, and we had all developed the "meteorite dismount" reflex. However, even with all our combined experience, it was not that uncommon to stumble across an occasional rock that might be a meteorite or might not. These rocks potentially represent the most exciting finds of the season since they may represent a new and unique type of meteorite. Or they may just be coal, basalt, weathered dolorite, or some other type of dark junk rock. Now in the most conservative of all worlds, you should simply collect every rock that has a chance of being a meteorite, the theory being that it's better for the curators at NASA's Johnson Space Center in Houston, the ultimate home of the collected meteorites, to sort it all out. After all, they have access to a host of analytical techniques to help them make a decision, they have plenty of time to consider the issue, and they are *warm*. However, in practice this does not always happen. In order to properly collect a meteorite, you must photograph it, wrap it with an identifier tag in special bags using a special teflon tape, record the particulars of the find in your field notebook, and so on. These activities require that you remove your gloves, and while the people in Houston may be warm, you most certainly are not. In the interests of comfort, and sometimes safety, searchers are reluctant to remove their gloves unless they think they have a real meteorite to deal with. Second, and perhaps more importantly, there is the issue of pride. The members of ANSMET expeditions are largely meteoriticists, that is, professional scientists who study meteorites. Now if you are a highly respected meteoriticist, you do not want to accidentally collect and bag a sample that ultimately turns out to be junk rock. Heaven forbid you should be exposed to any light-hearted ridicule should the folks in Houston discover you have shipped them a penguin dropping or some such item.

So, what happens when a party member finds a rock that might be a meteorite, but they're not sure? Generally, everybody gathers around the rock and argues about it. This usually solves the problem, either because some consensus is reached, or because someone finally decides it would be less painful to remove his or her gloves and collect the rock than to continue to listen to the discussion. Unfortunately, the discussion surrounding this particular rock failed to generate a consensus. I thought that the surface of the rock looked so much like a fusion crust that we should collect it, just in case. Randy and Robbie thought it was a piece of junk rock and pointed out the other terrestrial rocks that were nearby. Monica thought maybe we should pick it up, but didn't seem inclined to fight very strongly about it. And, of course, with the low temperatures and high winds of the day, *nobody* wanted to take their gloves off. Tired, cold, and frustrated by the day's events, Robbie and I both got stubborn, and the "discussion" reached an impasse.

Finally a compromise of sorts was reached. It was decided that I would put the rock, which weighed about one and a half pounds, in my pocket. If we didn't find any other rocks of this type for the rest of the day, we would assume that it wasn't junk rock, and I would bag it for Houston. If, on the other hand, we found many more rocks that looked the same, we would assume the rock was junk rock, and I'd pitch it. This verbal treaty endorsed by all, the discussion was closed, and we all grumpily went back to our snowmobiles and resumed the search sweep.

Unfortunately, as is often the case with treaties, things broke down. Later in the day Monica found a second, smaller rock that looked just like mine. The discussion was renewed. While we had agreed what we'd do if we found no more rocks like the one I'd found, and what we'd do if we found a large number of rocks like the one I'd found, we hadn't agreed what to do if we found only one more. By this time the cold and wind were taking a real toll, and this exacerbated the unresolved frustration from the previous discussion. The debate blossomed into something closer to an argument. Finally enough verbiage and vitriol were generated that I crossed the threshold where frostnipped fingers seemed a less painful alternative than continuing the discussion. I announced that I would bag my rock, and I didn't care what anyone else thought. This I promptly proceeded to do. Monica then decided that she had no alternative but to bag her rock, stating that she would never forgive herself if she pitched her rock and later found out that mine was a unique specimen. Both rocks were stored away and with a few curt exchanges, we returned to the search.

We ultimately collected twenty-four samples during the search, but the wind was brutal and work proved to be miserable. It became necessary to make frequent inspections of each others' faces to check for frostnipped noses, cheeks, and ears. Finally, cold, sore, tired, and still a bit miffed at each other, we decided to call off the search and return to camp early.

January 14 began well but after two hours we had our first problem. Monica's snowmobile threw out one of its rear axles. This had happened to us once before as the result of the snowmobile flipping on a steep slope. After the first time we had been able to force the axle back into position with a lot of grunting and banging of tools. Unfortunately, the majority of the expedition's tools were back at the main camp on the Lewis Cliff Ice Tongue. We decided to temporarily abandon the snowmobile and continue the search. Monica now had to sit on the sample box on the back of my snowmobile. We continued to find a few new meteorites but were ultimately forced to give up on the western portion of the ice field because a fresh snowfall during the "night" had covered the ice and hidden most of the meteorites.

At this point, with a further search to the west looking to be unproductive, Monica and Randy took his snowmobile back to her snowmobile, and they towed it back to camp. In the meantime, Robbie and I went up to the crest line of the field, where the fresh snow had already mostly blown away, and hunted for meteorites there. We had found a few meteorites when Robbie's snowmobile, affectionately known to all of us as "The Hog," stalled. It had been doing this on and off for a number of days, and we thought there might be a problem with the idle. In the past, it had always started up again after a few vigorous tugs on the starter lanyard. This time, however, it absolutely refused to start. Fearing that Randy and Monica might be getting worried about our continued absence, we returned to camp with Robbie riding on the back of my snowmobile.

At this point I was getting more than a little worried. Two of our four snowmobiles were incapacitated, and we needed to rendezvous with the other four members of the party back at the Lewis Cliff Ice Tongue in the next few days if we were to have time to traverse together to the abandoned Beardmore camp on the Bowden Neve for our pickup by a ski-equipped C-130 Hercules aircraft. It was getting very late in the season, the summer weather was already showing signs of breaking down, and there wasn't much of a margin for delay. While we could probably abandon one snowmobile and get back to the rest of our group with three snowmobiles, or perhaps even tow the fourth snowmobile behind one of the others, there was no way we could make it with just two snowmobiles without abandoning a lot of gear. While I wasn't worried about getting back safely, I didn't want the legacy of my first Antarctic command to be a pile of abandoned equipment at the MacAlpine Hills.

We discussed our options as we ate dinner in Monica and Robbie's tent and decided that the first order of business was to get Robbie's snowmobile back to camp. Then at least our problems would all be in one place. After dinner Randy and I, with Monica on the back of my snowmobile, drove back to Robbie's machine. The ice, with its dusting of new snow, now proved to be so slippery that it was difficult to stay on one's feet. After a lot of skating around and a few falls, we determined that the snowmobile still would not start. We decided to rope the two working snowmobiles in tandem and use them to tow the dead machine back to camp. Unfortunately, the slippery ice did not give the snowmobiles any purchase. Try as we might, we could not get Robbie's snowmobile up the rise toward camp. After a great deal of skidding, engine revving, and shouting, we gave up. This left us no choice but to go down. We reversed the whole arrangement and pulled the whole mess down the slope and away from camp until we reached the edge of the ice field. Our traction improved immensely once we got onto the firn and snow at the edge of the field, and at this point it was simply a matter of towing Robbie's machine the long way around to camp.

The day ended with another twenty-two meteorites in our collection box, but with two broken snowmobiles sitting outside my tent, I wasn't finding that to be much of a comfort. We had to be back to the Lewis Cliff Ice Tongue by the seventeenth, the eighteenth by the latest, if we were to have a chance of getting to the C-130 pickup point on time. I went to sleep still trying to decide what to do next.

The next morning we managed to establish reasonably good communications with the South Pole base for a short time. We also heard South Pole talking to the rest of our party at the Lewis Cliff Ice

Tongue, although we couldn't hear their response. Feeling somewhat heartened by being able to share our problems with the outside world, we ate a quick breakfast and decided to attack the snowmobile problem, beginning with Robbie's. Mercifully, the wind was not too bad. We traced the electrical system and didn't find any obvious problems. On a whim we tried switching the electronic starter with the unit from one of the working snowmobiles. This didn't help, so we switched them back. I pulled the plugs and determined that they were absolutely filthy. After putting in all new plugs we tried again. Still nothing. I pulled the carburetor cap and had Randy turn the engine over. While it was hard to tell in the wind, I couldn't smell gas. By removing various hoses in the fuel system and blowing and/or sucking on them, I managed to remove any potential blockages and get gas through the system (and also into my mouth). I reconnected the hoses and Randy gave the lanyard a yank. Robbie's machine, which had the reputation of being hard to start even when it was working properly, started instantly. My sense of relief was profound. Thank God, now nothing needs to be abandoned. One down, one to go.

The axle, axle bearings, and bearing holders on Monica's snowmobile's rear sprocket drive proved to be a shambles. The axle was badly scored, the bearing was frozen, and the bearing holder plate on the suspension was bent almost ninety degrees out of true. After a minor struggle, we managed to put the whole mess back together again, even though it looked like the axle would be turning against the bearing rather than the bearing turning itself. Since we were lacking most of our tools, I was forced to spend nearly half an hour drilling a hole through the snowmobile's chassis by twisting a drill bit back and forth with a pair of channel lock pliers. I was then able to use the hole to bolt a hinge from one of our food boxes on the inside of the chassis where it pressed against the side of the bearing holder. The result looked ludicrous, but it was the best we could do under the circumstances. Monica started it up and took it for a gentle trial spin. The bearing squealed. It smoked. It smelled. It got too hot to touch. It worked!

We were in business again! After a leisurely lunch we took a short test trip and searched the upper portion of the ice field near camp. We found only two more meteorites, but one was a rare type of primitive meteorite called a carbonaceous chondrite. We decided it would be best not to press our luck, and we returned to camp. In celebration, we broke out some of the last of the precious frozen lobster in the food rations. Over lobster and the first relaxed conversation in days, we decided to head back to the Lewis Cliff Ice Tongue on the seventeenth. That gave us one last day to search for meteorites. But for the search to be effective, we needed a good blow to clear the snow off the ice field so we could see the meteorites better.

Alas, it was not to be. We woke to find that even more fresh snow had accumulated while we slept. Not only was meteorite hunting going to be next to impossible, but we were now in serious danger of losing sight of the snowmobile tracks we had made when we traversed in from the Lewis Cliff Ice Tongue. If our old tracks were covered, it would be considerably more difficult to avoid intervening crevasse fields and the cliff itself as we navigated back to the rest of the party. After breakfast we tried the radio and were surprised to immediately raise the South Pole station radio operator. He managed to contact the other half of our group and then served as a relay so that we could have a quick conversation. Given the lateness of the season and the fresh snow cover, it was decided that it would be best if we returned to the rest of the party immediately rather than the next day.

Curiously, despite all the excitement involved, we ultimately didn't share many of the details of our trip with the rest of the party. However, I later noted that the rocks that Monica and I had found, and over which we had argued so vehemently, had been listed in the official logs as probable terrestrial samples, that is, junk rock. This irritated me considerably, but after some thought I decided not to raise the issue. So, with a conscious decision to put the MacAlpine Hills trip behind me, I joined in with the rest of the team and started preparing for our traverse to the abandoned Beardmore Camp and our ultimate extraction from the plateau.

It wasn't until six months later that I was to think about the MacAlpine Hills trip again. By this time the expedition had already been driven from the forefront of my attention by the details of my job at NASA's Ames Research Center. Then one day the phone rang. On the other end was one of the staff members at the meteorite curatorial facility at NASA's Johnson Space Center where the meteorites recovered from Antarctica are taken, stored, and given their preliminary examinations. They just wanted to let me know that the two pieces of junk rock that we had argued so energetically over at the MacAlpine Hills had turned out to be *lunar* meteorites. Yes, pieces of the moon! The one I had found was at least ten times bigger than any of the six previously known lunar meteorites and currently remains the biggest one ever found.

To this day I have a hard time grasping this. To think that, at some point in the distant past, probably within the last ten million years or so, a comet or asteroid hit the moon and flung a piece of lunar rock into space where, after many millions of years, it fell to earth in Antarctica. It then resided in the ice for thousands of years before spending one day in my pocket and moving on to the Johnson Space Center. I find it difficult to integrate activities played out on such a cosmic scale into the everyday workings of my life. Still, I guess it is possible to gather at least one moral from the story: "How many penguin droppings are worth one lunar meteorite?"

***Scott Sandford is currently working at NASA on issues associated with the formation of new planetary systems, astrochemistry, and the origin of life. He grew up in Los Alamos, New Mexico, surrounded by scientists.***

# Solo

SUSAN L. ADKINS

I follow the yellow line. Six months ago, I trailed from one side of the taxiway to the other, and I chased this wide stripe as though it was moving erratically instead of me. Often I drifted out in a wide arch over the dark asphalt before correcting back toward the center. Then each press against the rudder pedals was either too much or too little. Today my feet finally know how to play one pedal against the other. Though calloused and laced up in thick shoes, they feel new and agile—a slight press of the left sole to avoid a patch of crumbling pavement, now a hard push against the right to make a sharp turn to face into the wind.

I am taking my time, though, going slowly, following the checklist, thinking before I key the microphone to talk to the tower. The oil temperature and pressure look good, rpm checks out, engine sounds right, smells right. There's a light perfume of 100 low-lead aviation fuel in the cockpit, but that is from my clothes. I was taught not to trust fuel gauges, so I manually measure the tanks before every flight, using a hollow, plastic dipstick calibrated from top to bottom. It's the old straw-in-the-drink trick we always liked to play as children. Insert tube into the wing tank until it touches bottom. Place finger firmly over top opening, remove tube and read usable gallons, stuff gauge in jeans' pocket, smell like avgas the rest of the day.

I am surprised that I have come to love a petroleum smell, but aviation fuel smells sweet—not pungent like automobile gasoline. Refiners add a light blue dye to 100 low-lead avgas to help distinguish it in color from other grades of fuel, but the characteristic aroma results purely from the refining process. Days after a flight when I empty the dirty clothes basket onto the floor for sorting, the smell wafts up from the heap, and I am back at the controls of a Cessna 152. Writing in *A Natural History of the Senses,* Diane Ackerman calls this a landmine: "Hit a tripwire of smell, and memories explode all at once." One whiff of aviation fuel brings back the thunder of the engine, the press of the headset around my ears, the resistance of the yoke against my left hand, the friction of rubber tires gripping asphalt runway, the release from earth, the climb, the wind that sometimes bumps, the wind that always grabs machine and me with it and finally lifts.

The earth recedes and flattens out beneath. Just a few hundred feet up, I can see the whole of Virginia's lower peninsula. When I first started taking flying lessons, I never saw this landscape. My hands were glued to the yoke and throttle while my eyes fixated on the instruments. About the third flight, I finally looked outside and fell in love. Twelve miles at its widest point, the peninsula is surrounded by the Chesapeake Bay and the James and York rivers. I was shocked at how small and immensely beautiful it looks from the air.

From only a thousand feet up, it is a green spit of land lying in a mass of blue water. Bay tributaries with Anglo-Saxon and Native American names like Newmarket Creek and Poquoson River cut into its edges. To the southeast is the geometric landscape of industry. To the northwest, hardwood and pine forests still cover the land, defying my perception of the whole area as one giant knot of parking lots and housing projects. Bridges project from its shore, and as the plane climbs higher, these giant structures shrink into thin, gray lines drawn across wide tidal waters. The peninsula's profile is so easy to read I often have to remind myself that it is not just a computer simulation on the local weather channel.

Today I am anxious to see it all again. The tower calls my airplane's tail numbers. "Seven-quebec-golf cleared for takeoff . . . winds zero-seven-zero at 8 knots . . . altimeter two-niner-niner-seven." The controller runs the wind direction, speed, and altimeter setting together, each word rushing to overtake the next.

My right hand flips on a navigation light. "Seven-quebec-golf cleared for takeoff," I echo back as I press the fuel mixture control to the full rich position. A quick taxi and I am ready on the runway, rolling down the center line, pushing the throttle to the max, steering with my feet, pulling back the yoke, launching, giving lots of right rudder. I am thinking, reviewing, trying not to forget anything, trying not to end up on the front page of the newspaper. "Local Woman Flies Cessna Into F-15, Costs Government Millions" . . . "Local Woman Forgets to Check Fuel, Runs Out on Takeoff" . . . "Local Middle-Aged Woman Should Stay Home." It is my first solo cross-country, and I miss my instructor.

Climbing to 2,500 feet, the aloneness is overwhelming. "Checklist, checklist." I call the words out loud. "Yes, talk out loud. Go slowly. Ask a question. Answer it. What do you need to do next? Take your time. You have all the time you need. No need to rush. Look outside for traffic. Think. What's your heading? Right. Now checklist. Checklist."

The Newport News shipyard and the marine terminal mass below me. Save for the sycamores and maples and crepe myrtles that line the old neighborhoods, every square foot of land is covered with something man-made. Crisscrossed with telephone poles, streets, houses, fences, dry docks, piers, mechanical cranes, and giant mounds of coal, a makeshift runway is nowhere in sight. Even the few ballfields in this East End section of town are bordered by drainage ditches that look dark and deep in the low morning sun. They could easily swallow a two-seater plane like this one, coming in for an emergency landing. I spot an open patch of grass beside the railroad tracks that split the peninsula down its twenty-five-mile length. Could I miss the power lines further up and land on that strip of earth if I had to?

"Seven-quebec-golf over the shipyard. Clear to the south." These are my last words to the Langley tower for the next few hours and the last before confronting solitude. I have made several short flights by myself, so the loneliness is not a shock. Still I do not understand its strength. I am practiced at being "one." I make long road trips by myself, I dine alone, watch movies alone, live alone. Loneliness is chronic. My friends call it the middle-aged divorcée's arthritis—a red swollen ache inside that promises to grow worse during the coming years.

The loneliness of solo flight is different. It is not within me. Rather, it materializes from the emptiness of the cockpit. I let my mind give it a presence and even a shape to make it less threatening. After watching Star Trek all my life, it seems natural to animate my fears. From out of the corner of my eye, I watch it stretch out on the right seat beside me. Its face opens in a wide hollow smile and mouths words I cannot hear. The unmuffled scream of the Cessna's engine saturates the air, and the monotonous drone swells into a white noise that absorbs all others. Below, a train moves toward the coal piers, but there is no sound of brakes scraping against steel. With my ears sealed beneath my headset, the earth seems mute and distant. Aloneness sits beside me, and I imagine its thin lips slowly puckering, its cheeks pulling back to call to me. I read the shape of two words: "You're—it." Yes, I am "it." I am pilot in command. I am alone at three thousand feet above the surface of home and still climbing.

My fellow student pilots talk about this loneliness too. One has decided it is merely a defense mechanism of her mind, designed to make sure she stays alert. Another is convinced his psyche is throwing up distractions, bent on self-destruction. I wonder if it is something unleashed by altitude alone, something that strips away all our pretense. On the ground, there are so many people around at work, so much traffic, so many e-mail and voice-mail messages, it is easy to ignore the fact that I am alone. However, when I confront my aloneness at altitude, I must accept my oneness. There is no avoiding my separate, solo self. I look over at the right seat one more time, kick the smiling remnant of my imagination out the right side of the plane, and fly on.

The James River is at its widest here as it passes the shipyard—more than five miles across to what is known as the Southside of Virginia. This river is a liquid chameleon that changes with the sky, the light, the season of the year, the time of day. On cold January afternoons when the sky is dry and cloudless and the sun still high, the James is as easy to read as a color-coded navigational chart. From a few thousand feet in altitude the variations in the depth shade from light to dark and in hues that range from green and blue to brown. The shallows are startling, for most of the river here is only five to ten feet deep. The channels run dark, though—some merely twenty feet in depth while the trench by the shipyard digs seventy feet beneath the surface. Topsoil eroding from a freshly turned Southside cotton field is a bloody plume contaminating the greenish blue of the river. It is all sharp colors and fractal shapes—unseen except by a lucky few.

This September morning the surface hides everything. The sun is still low, and the atmosphere is heavy with humidity. As the earth warms, a layer of haze will grow from the moisture and smoke and dirt suspended in the air. For now though, the river rivals the sun. It doesn't matter that its surface is roughened by the wind. It reflects everything, radiates light in every direction, blinds me when I look down. The James appears hard, set in concrete and mirrors. It has lost its blues and greens and bounces metallic light from its impenetrable surface.

I was more than forty years old the first time I saw the Earth from a small, slow flying aircraft just a few thousand feet in altitude. For the first time in decades, I could not automatically attach names to everything I saw. The high angle of the light that day played with my perception, reducing the world from three dimensions to one. Height and depth blended into width. Green and black spots dotted the edge of a narrow road below. They were round and flat and lying abstract against the light gray of winter grass. Only after several seconds of dissection did I separate a green tree canopy from its twin black shadow. My middle-aged eyes were like those of a child again, learning to read and interpret.

For a new pilot at least, it is easy to be hypnotized by the Earth. Its patterns of color and shape become sirens, luring me away from checklists, fixing my attention to a single point on the surface, making me forget to watch the heading indicator, coaxing me to lose my way. I cross the James, heading almost due south en route to Elizabeth City, North Carolina. To the east lie the piers of the Norfolk Naval Base. To the west, the open fields of Isle of Wight County are reassuring. If the engine quit right now I could glide there, just pitch the nose to maintain 65 knots, circle round one field, add flaps, give right rudder against left aileron to slip the plane down past the barrier of trees, flair just above the rows of peanuts, touch down in the sandy soil. I tell myself I could do this if I had to.

Leveling off at 3,500 feet, I scan the instrument panel: air speed indicator, vacuum gauge, altimeter, tachometer, ammeter, fuel gauges, engine instruments. I lean the fuel, adjust the throttle, and the air speed indicator climbs to 110 knots. I have climbed out over six miles in eight minutes and am just two minutes behind my projected schedule. My first stop is 45 miles away. If the winds aloft hold true to the weather report, this should be a short twenty-five-minute flight.

I like navigating above the Tidewater area. The rivers carve at the land, washing wide in some areas and narrow in others. Each bay has a distinctive outline. Every inlet cuts through the marsh in a unique way. It is easy to match features drawn on the map with the real ones on the ground. Both on paper and on land, the Nansemond River twists its way south and narrows into a channel that meanders through a wide marsh. I keep on my southerly heading, overflying a small grove of radio towers that stand more than one thousand feet in height.

The first time I ever flew over these spires I looked directly down one, and the ground rushed right up to me. For an instant, I might as well have been balancing on one foot atop the pinnacle of a skyscraper. The whole picture was different. The abstract numbers on the altimeter suddenly transformed into a concrete link from me to the earth. My mind measured the distance and banged out the numbers in my stomach while it sent my head on a roller coaster ride. For a few seconds my perspective shifted back to three dimensions and my mind screamed, "Fall!" A look toward the horizon separated me from this virtual tether and put my stomach and head back in straight and level flight. "You looked, didn't you," my instructor teased. "I told you not to look down on those things."

The dark green of the Great Dismal Swamp surrounds Lake Drummond, a body of fresh water two miles wide and almost three miles long. Amidst the vegetation of the swamp, this lake is an easy checkpoint for all pilots entering or leaving the Norfolk area. I keep this opal of blue well to my right as I head south, watching for other landmarks. Inside the cockpit, I also eye the steady white needle of the VOR indicator. These three letters are short for Very high frequency Omnidirectional Range system. A VOR station at the Elizabeth City airport transmits radio beams in all directions. I track on one of these radials, keeping the needle on my indicator centered. Between following my magnetic heading, tracking the VOR signal, and checking outside landmarks against those drawn on the sectional map, I know exactly where I am.

Ten miles out of Elizabeth City, I call the tower. "In-bound for touch-and-go." A woman's voice answers back, reporting the wind conditions and the current altimeter setting. "Expect runway two eight," she says. Immediately I glance at the north, south, east, and west markings on the heading indicator. As I check its alignment with the compass, I start sketching the picture in my head. To help pilots land on the correct runway, each is labeled with numbers that correspond to the compass. Runway two eight's magnetic direction is

280—almost due west. If there is more than one runway at an airport, the second and third are laid out to align with other common wind patterns.

I descend to a thousand feet and enter the traffic pattern—a three-sided box flown around the end of the runway. First I fly with the wind, then turn perpendicular to it before finally flying into the wind for the touchdown. I talk to myself a lot now. It's one command after another. Make sure the mixture is rich, carb heat on. Now power back, lower flaps, slow the plane down. Look outside, look inside, now back out. Fly the box. Watch your position. Check your altitude. Scan the instruments.

"Seven-quebec-golf, cleared touch and go, runway two eight." The woman's flat Carolina vowels sound ten degrees more southern than my own Virginia-trained diphthongs.

Turn onto the final approach, down to five hundred feet now, keep descending on course, add the last notch of flaps, slow to 60 knots, read the numbers *2* and *8* on the end of the asphalt. "Thank God," I mumble out loud. Out of a possible four runways, I've managed to pick the right one without my instructor's help.

Keep the plane centered now. Power back to idle. Over the numbers, pull back slightly on the yoke. Level off for a few seconds. Gently now, pull back more, rotate the nose up, slow it down, settle to the surface, touch first on the two main wheels, finally lower the nose, let it roll, but don't relax. Flaps up, throttle to full power. Just touch rubber to asphalt and take off again. That's all the FAA regulations require for this part of my cross-country.

I am on the go, tracking the center line, pulling back on the yoke, lifting off at 55 knots, climbing out at 67 knots, heading toward the northwest on the second leg of my triangular route. Seventy miles away and just across the Virginia state line lies a small town called Emporia with an equally small untowered airport. With this headwind it will take fifty minutes to reach. Once Elizabeth City cuts me loose, I change the radio to the common frequency for Emporia.

It's probably too far out to hear anything, but I listen for traffic while studying the farmland below.

Like eastern Virginia, this Carolina terrain is flat, but the wide tidal rivers are missing. For about twenty miles there is nothing but rectangular fields with precise right angles. Tracks of yellow tobacco stalks alternate with acres of freshly turned soil. The farms line up like one emergency landing strip after another, all pointing in the same direction. Such a well-ordered landscape is rare in the east where highways usually snake around creeks and follow the meandering trails of long-forgotten cow paths.

This parquet land disappears in a few minutes, fanning out into woods and pastures and rambling roads. A few houses, a couple of gray barns and a filling station cluster at almost every crossroads. I look for the one that the map labels Gatesville—a dot where two roads merge into one. A small lake is drawn off to the right of the village, and the giant blue tail of the Chowan River is sketched about four miles to the left. Though less than half a mile across, the Chowan is impossible to miss or confuse with anything else. However, the rest of the terrain is nondescript—pasture and farmland webbed together with little two-lane roads joining and dividing and rejoining again. Tiny villages populate the landscape at random—most looking like miniatures cast from the same mold.

I check my heading for the first time since leaving Elizabeth City's air space. Though the indicator shows I'm ten degrees off my projected course, I instantly discount it. After all, I see the river. "There's no mistaking the Chowan." I whisper the words out loud. The intercom captures my voice, amplifies its timbre, swells its pitch and pipes the words through the tiny speaker in my headset. Electronics inflate you with the richness and authority of your new voice. Alone with this microphone and no chance of anyone ever knowing, I can't resist the idea of trying out my James Brown impression.

"Wo-o-o-o-w, I feel good. Dada-dada-dada-da. Like I knew that I would." Everyone has these totally narcissistic moments, I tell myself.

You've done something that was hard—really hard for you at least. Give your ego free reign and suddenly genius seems not only possible but probable. Without a doubt you sound great, and God knows, you must look good too. "Dada-dada-dada-da! So good, so good, bum, bum, bu-m-m-m!"

Learning to fly was hard for me. Having to change instructors twice forced me to readjust to teaching styles every few months. Then there was a long stretch of icy weather that left me grounded weeks between flights. However, most of the time, I like to blame the outlandish amount of information that's involved. From the ground, the sky may look open and limitless, but actually it's carved up into different kinds of airspace with lots of regulations governing their use. From learning how to communicate with the tower to understanding weather, right-of-way rules, navigation, and medical requirements, there are thousands of new facts to learn and remember.

In addition to the cognitive load, the body itself has a physical part to learn. A student pilot can read a step-by-step description of how to land a plane. She can chair-fly at her dining room table and practice the steps every night for a week. Both help, but the dining room has no weather, and the chair has no wings. With how much force do your legs push against the rudder pedals when there is a cross wind? What does the engine sound like when you pull back the throttle? Are you supposed to feel like your bottom is being nailed to the seat when you do a steep turn? Without experience the body has no memories to help gauge its reactions.

A three-pound brain sits atop our bodies, packed tightly with about a hundred billion nerve cells that biologists call neurons. Within this mass of highly structured, wiry protoplasm, learning happens. Camouflaged behind gray folds of tissue, scientists are still trying to map its complex geography after a hundred years of study. Researchers at the University of Iowa have used an MRI scanner to discover that one area of the brain processes the names of people, animals, and objects while a different section coordinates verbs. Yet a third region is required to converge nouns and verbs into sentences.

Neurobiologists tend to be as highly specialized in their research as the brain is in its discrete parts. Some only study the mechanics of motor-skill memories and the signaling system that enables tasks like walking and running to become automatic. Dangling from the edges of each neuron are fibrous dendrites that receive incoming messages. There is also an axon fiber that sends the message on to the next set of neurons. A minute gap known as a synapse exists between each cell so that they never even touch one another. Thus messages are distributed using two types of signaling systems. A message is transported across the body of an individual cell and along its axon as an electrical impulse. However, the synapse itself is bridged chemically with molecules of a transmitter substance flowing from one neuron to the next. When I practice my landings, I often imagine the synapses in my head growing wider with each climb out. Once I half-heartedly tried to convince my instructor that after six or seven landings the gaps had grown so large that the neurons could no longer transmit from one to the other. "Nice try," he said with an amused look. "Let's go round again."

I tried to figure out the reasons behind my slow progress—read a lot and even interviewed a few experts studying the learning process. When I talked with Dr. Reza Shadmehr, assistant professor of medicine and biomedical engineering at Johns Hopkins University, he recognized the situation. "When people are just trying to learn a task, they have trouble doing anything else. For example, if they try to carry a cognitive load . . . while trying to learn a new motor skill, the learning will be slower." Sometimes existing skills that "require your limb to be used in a very different way will interfere with learning a new skill. If someone had learned flying without having learned to drive, they would not have that interference. . . . You do worse because you've learned a skill that was somehow anti-correlated to this task you're now doing. Eventually you might learn . . . and keep both skills."

Shadmehr and his research associates recently discovered that "when you learn something new, the brain requires time for it to consolidate . . . it needs a period of off-line practice." No one knows exactly how much time the brain needs, but current studies suggest it may be five to six hours. Once this time has passed, the brain is able to handle distractions and carry on with other activities while performing the task.

Using positron emission tomography (PET), Shadmehr monitored the changes in blood flow inside the brains of participants who were learning a new motor skill. The PET images showed that areas on the top and side of the brain are most active when the skill is first being learned. After five or six hours away from the task, the subjects were allowed to practice the skill again. This time the blood-flow activity shifted from the front of the brain to the back where the cerebellum is located. Apparently, even when the task is no longer being practiced, the brain is still working in the background, processing the information. Once the brain has time to consolidate the new information and move it to the cerebellum, the information about the skill becomes stable and is resistant to interference or change. This may help explain why motor skills that are learned incorrectly are extremely difficult to relearn.

Shadmehr believes his research "supports the idea that the brain creates a blueprint, or model, for performing a particular task." Once the blueprint is in place, the task becomes automatic. However, if a second motor skill is introduced within a couple of hours after the first, the brain is not able to reorganize and store the information in the cerebellum. When this happens, the simple motor skill requires more time and practice to become automatic. If the brain is also trying to recall FAA regulations, procedure requirements, aircraft systems, and fine motor skills while coping with new physical sensations, it's easy to imagine the cranium afire with all the reorganizing and imprinting that must be going on inside.

Above the monotonous landscape of eastern Carolina, I decide to give up on finding Gatesville and its lake, and I start looking for the next place in the road. When the Chowan forks, I stick close and parallel its western branch. Murfreesboro is a little larger. That's got to be it right there. That's got to be the trailer park shown on the map, but should it look so close? It's supposed to be at least five miles south of my path, but it's almost directly under me. From 4,500 feet, how close should things look? I've only been this high a few times, and I can't remember the picture.

The VOR's needle is deflected as far as possible to the right of center. I switch to the Coefield VOR station, which is much closer, turn the dial until the needle centers on a radial, pinpointing my location. "That can't be right." I holler the words, and this time my voice startles me as it blasts back through my headset. I switch to Franklin, center the needle and again don't believe what I read. Triangulating off these stations shows me south and east of where I think I am. I realign the heading indicator with the compass. If I follow what it shows, I've got to make a big turn to the right, but the landmarks are right below.

In *West with the Night,* Beryl Markham wrote, "A map in the hands of a pilot is a testimony of one man's faith in other men. . . . A map says to you, 'Read me carefully, follow me closely, doubt me not. . . . Without me, you are alone and lost.'" I study the map and the terrain for another fifteen minutes and then start to talk to myself again. Could that clearing way out there be the airport? The sectional map shows the airport east of town and right next to Route 58, a four-lane road. Don't see the town yet, but that's gotta be the airport, so I start my descent.

"Emporia Unicom, Cessna seven-quebec-golf ten miles southeast of the airport. Inbound for touch-and-go. Request winds and runway advisory." There is no answer, but that is not a surprise. At many untowered airports, the manager has to fuel airplanes and sell candy bars behind the counter as well as man the radio. So I call the other airplanes that might be in the vicinity, but there is still no response.

I'll overfly the runway at two thousand feet, take a look at the wind sock to see what direction the wind is blowing, look for other traffic, then circle back and descend for my landing. But as I approach the airport, it disappears like a mirage. Instead of dark pavement and hangars, there's a grove of evergreens growing in the middle of a pasture. I see a four-lane highway and easily convince myself that it is Route 58. No problem. Just follow this, and it'll lead you right to the airport. Be patient. You'll see it in a minute.

Smokestacks and plumes of exhaust appear out of the haze. There's a town, but there's no airport in sight. I recognize I-95 and wonder why the map shows it to the west of town when it obviously is to the east. The heading indicator shows I'm traveling southwest, but something doesn't fire in my brain so I ignore it. It smells like a pulp-mill town, has a sizable river cutting through its middle, and doesn't look a thing like Emporia. Finally, my brain flares. I'm lost. I didn't ignore the VOR, the heading indicator, the compass, or even the map. I simply failed to believe any of them. When the landmarks didn't match the map, I convinced myself they were close enough. For the last forty-five minutes, I've been discarding square data through the round holes in my judgment.

Training kicks in again, and I climb and circle the area, trying to get a fix. I unfold the map, stretch it from window to window and study the draftsman's artistry—magenta highways, black railroads, blue rivers, green landscape. My eyes strain and I feel my gaze skimming across the map like nervous fingers. If I could pull over to the side of the road, I could figure this out.

I dial in the radio frequency for Washington Center. "Cessna seven-quebec-golf, student pilot, at three thousand feet. I'm lost. Trying to go from Elizabeth City to Emporia." A man's voice replies.

"Cessna seven-quebec-golf, squawk two-four-two-six and indent." I echo back the numbers and then dial them into my transponder before pressing the indent button. "Give me a minute to find you quebec-golf. Be right back with ya." Geez, I thought when I hit the indent button my position would light up on the radar screen like an atomic bomb.

I scan the area for traffic then take another quick look at the map. There it is. A four-lane road, an interstate, a river and a sizable town. I'm over Roanoke Rapids, North Carolina. "Seven-quebec-golf, we've gotch'ya on radar. You're over Roanoke Rapids. You're only about twenty miles south of your destination. Turn right to a heading of zero-four-zero. We'll stick with ya the rest of the way. Let me know when you have the runway in sight."

I watch the heading indicator and scan the horizon for Emporia. Silently it emerges from the layer of haze hugging the ground. Everything is in the right place—the highway, the town, the three runways that look like a big capital "A" drawn with gray asphalt. The place is surprisingly quiet on this sunny day. No traffic at all. I make the radio calls, ease the throttle back, add flaps, fly the box, cross the numbers, touch-and-go, add full power, switch flaps up, pull back on the yoke. I feel my pulse push against the side of my neck—a signal to relax. My shoulders are practically hugging my ears, and my arms must each weigh fifty pounds. I sit back, take a deep breath, and with it, a wave of relaxation rolls up from my toes and out through my fingers. My right hand automatically reaches to adjust the trim, inching the nose up. I climb out, turn east, and head for home. The sun is high overhead. The surface temperature is in the upper eighties, but I'm cool and comfortable at 3,500 feet. "I feel good." My voice sings out strong and clear. "Dada-dada-dada-da."

The liberating thing about flying is that you're not expected to be perfect. Flying instructors don't fail you just because you make a mistake. In fact, they give you many opportunities to blunder, and then they train you how to avoid mistakes and the behavior that leads to them. Much of this training concentrates on self-awareness of your personal limitations and the perils associated with both the absence and overabundance of confidence. Also, imprinting through repetition is one of the touchstones of aviation instruction. Do it a thousand times, and then do it again. Practice landings like James Galway on his flute. He's practiced so much that when he reads each musical note printed on the white page, his fingers, diaphragm, tongue and lips automatically shape to create the sound. Through practice, the memory becomes visceral, functioning as though it resides within the muscle tissue itself.

Memory aids often take the form of checklists that are faded and worn by repeated use, even by pilots who have been flying the same aircraft for the last twenty years. They also show up in easy-to-remember acronyms and phrases. Do your GUMPs check before you land; that is, check your *g*as, *u*ndercarriage, *m*ixture, and *p*rop settings. When you get lost, remember the "Four C's"—climb, communicate, confess, and comply. Climb for altitude that will increase your range for radio reception and radar detection. Communicate with any available facility, and confess what your predicament is. Finally, comply—do what the controller tells you to.

I invested many hours in mastering this machine built especially to harness the elements. All the while there was a kind of alchemy at work: altitude became the catalyst for a change in perspective. When I was sixteen, I made my first road trip. My cousin and I drove thirty miles to the next town down the road and were astounded to discover we could do something without an adult at hand. Now today, thirty years later, there are still more lessons to learn. It seems to me that flying provides a good model for life—one I wish I'd learned when I was sixteen. Mistakes must be expected and even welcomed for what they teach us.

The James River materializes from the thick haze and stretches along the horizon, reminding me of Markham again. When she learned to fly, she claimed she also "learned to wander . . . learned what every dreaming child needs to know—that no horizon is so far that you cannot get above it or beyond it." I am an older child now, finally learning to wander. I look down on my home and realize how small my worldly life is. The house where I live is difficult to pick out from the air. The house where my former husband still lives is even harder to spot from above. The small stand of loblolly pines has grown up and crowned the whole yard with a dark green canopy. Children might have played there once, but from the air you can't see the missing toys. The hospital where I had my cancer surgery is gray brick and does stand out. Still, someone flying overhead who doesn't know the area would not be able to distinguish it from a school or an office building. If I had known all those years ago that it's not only possible but all right to get above things, climb high, to confess and ask for help, would it have made a difference? If I had picked up a map along the way, would I have read it? Would I have even believed what it said?

I touch each line on the descent checklist, gradually working my thumb from one item to the next. Mixture rich, power back, carb heat on, brakes tested, fuel checked, seat belt fastened, checklist stored. From my cockpit, the peninsula is a beautiful, silent spit of land floating in a mass of calm, blue water. One neighborhood links to another. Streets stretch out in a complex grid, but I don't have to follow any of them. I take one last look for today, trying to imprint in my soul the image of the peninsula and the horizon beyond. I want to recall this perspective at will.

"Langley Tower, Cessna seven-quebec-golf, over the shipyard, inbound for full stop."

*Susan L. Adkins was a long-time resident of Hampton, Virginia, who had many and varied interests. Her two greatest passions were writing and flying. She died at age forty-nine in a flying accident on March 1, 1998.*

# In the Dark

GEOFFREY ALEXANDER

Like most people, I assume, I have always hated hospitals. Can't stand the necessarily desensitized staff, the terrible, bottom-of-a-Lysol-can smell, the institutional hopelessness. I suppose that I have been fortunate in the manner of my relatives' passing; that is, my family members have always gone quickly into that good night, thereby sparing the rest of us the crushing experience of the interminable death watch.

I had never, before this year, had to spend any length of time in a hospital environment until I started a job as a telecommunications technician building computer cabling systems called LANs, or Local Area Networks. For those not familiar with the acronym, LANs are the cabling systems that allow computers to talk to each other directly, without phone lines. The lines can be copper twisted in a special manner to facilitate the quality of high-speed communication or, fastest of all, fiber optics: glass wire that carries pulses of light rather than electricity.

Generally my company installed these systems in places like banks or data processing centers, so I was more than a little surprised when I was assigned to a two-month-long project at one of the largest hospitals in San Francisco. The job required that my crew run cable through six floors containing laboratories, a blood bank, a cancer ward, and an entire floor devoted to children with diseases that I would rather not even begin to consider. Of course life and work often conspire to put us in those places we would most like to avoid. And, while I was reminded daily of my good health by barefoot children pushing rickety IV stands accompanied by doctors in white coats with beepers that played light-hearted tunes, while I nonchalantly squeezed to the back of the freight elevator to make room for the occasional body bag on a gurney, while I even learned not to flinch when, upon turning a corner, I bumped into a man with a ragged hole in his face where a nose should have been, perhaps it is indicative of human, or just my own, selfishness, in the midst of all of this suffering: I could only think about how much I hated my job.

I didn't hate my job for the usual reasons. Although I had gone through graduate school with a goal of working in publishing, I didn't hate it because I had failed to find a position and was now carrying a tool belt to work every morning instead of a briefcase. I didn't hate it because professional people (like doctors and researchers) often treat blue-collar workers shabbily. I didn't hate it because it was physically grueling, dirty, low paying, and occasionally dangerous. These things made me unhappy, but they didn't make me hate it from the depths of my chew-a-leg-off-to-escape animal soul. No, the reason I woke up every morning clenching my teeth was that the job itself scared the hell out of me.

Phobias, like any nightmare, spring from deep within our psyches. We know that deep down there is an explanation for seemingly inexplicable terrors. But, understanding the original catalyst of a phobia or nightmare is as worthless at 3 A.M. in a sweat-drenched bed as it is in seat 13F while one tries to block out the building roar of jet engines being run up for takeoff. Understanding that the boogyman is as likely to crawl out from under your bed as is Mister Rogers is as worthless as knowing the deaths per hundred thousand by airplane crash. In fact, the more one knows about such things, at least in my experience, the more one knows fear.

I am not by nature a fearful person. I live two miles from a major

California fault line in an area that burned to the ground three years ago. I regularly drive over the speed limit, though not excessively so, consume foods and liquids known to cause cancer, and engage in sports that have almost cost me my life. I have never felt compelled to dwell on death. I had no preparation for the uncontrollable panic of my first phobic attack.

Understand this: the job itself was a recipe for fear. All of the ingredients that might trigger a physiological, fight-or-flight response were there. I worked in the spaces between ceiling and floor in a twenty-two-story building where, in the very act of walking through the front door, I felt uncomfortable.

The cabling system in a tall building is not unlike a human nervous system. In the hospital where I worked, for example, there was a main riser that ran inside a shaft from the basement to the top floors. A riser is a wrapped group of cables, much like a spinal column, that branches out at each floor and connects to the hundreds of workstations that function like two-way nerves carrying information in and out of the network. With the speed at which technology changes, it is cyclically necessary that the cabling systems be replaced to measure up to the speed of the workstations. If it were possible to upgrade a human brain to a faster model, for example, it wouldn't do much good if it was still relying on an antiquated, plodding, nervous system.

In order that the opthamologist's brand-new, twenty-inch, high-resolution monitor be able to receive the information from its server quickly enough to accurately generate a three-dimensional modeling of your eye before laser surgery, somebody earns eight dollars an hour to climb up into a ceiling and pull in the cable for the system's upgrade. Somebody like me. In a typical day I spent up to six hours in an area ranging from twelve to thirty-six inches in height. An area so small that I often couldn't get up on my hands and knees. Imagine it. The space between each floor is jammed with a tangle of copper plumbing for water, sheet metal ducting for climate control, electrical conduit, cable, wire hangers from which the acoustic tile ceiling hung and the skeleton of the building itself: mighty girders covered in a gray, fluffy, blown fiberglass material that protected the girders against corrosion. This material itched like hell if it got on your skin, which it always did. Down the shirt, on the skin, in the eyes. One of my coworkers became convinced that God had built a special little corner in hell covered with the stuff for the worst sinners. And in the summer, the ceiling is hot.

We wore disposable protective suits made of a tough, paper-nylon blend that protected us from the fiberglass and dirt but also acted like a personal sauna. The suits had floppy gray feet and peaked hoods, so we called them bunny suits. We wore OSHA-approved gas masks because we didn't believe the hospital administrators when they told us the fungus that grew in the ceiling and was potentially deadly for immuno-suppressed patients wouldn't harm us in the least. My mask was rated for the following: "Dusts, Fumes, Mists, Radionucleides, Radon Daughters, and Asbestos Containing Dusts and Mists." Like soldiers in the movies, we moved everywhere in the ceiling on our elbows, bellies, and knees and, lest we put our weight on the wrong part of the ceiling and literally fall through into the room below, we could only move across solid, metal objects. If you have ever had to spread-eagle your body across a few lengths of crisscrossed half-inch pipe, you'll know that it can get awfully painful. I often went home looking like I'd been worked over by a bar full of bikers, with bruises and abrasions on my knees, chest, and elbows. As I think about being up there, my heart begins to pound.

The day I learned my mental limit marked a record breaker of a heat wave in San Francisco. Outside, the doctors in their white coats and nurses in their white shoes crossed the street, dully shining and listless as the paralyzed California flag that hung from its pole as still as drapery. I was working near the blood bank where the techs took any excuse to open the refrigerators stuffed with bagged liters of chilled hemoglobin and fan themselves with cool blasts of air. I donned my bunny suit, mask, and gloves, grabbed a flashlight and walkie-talkie and agreed with my partner as to the route I would pull the cable through the ceiling.

It was easily ninety degrees in the building; I was sweating before I even climbed the ladder through the access hatch. Up in the ceiling, where the heat is greater due to the hot-water pipes and lack of ventilation, it was one hundred degrees or better. The bunny suit bumped the temperature another ten degrees. I have since learned that extreme heat is a common trigger for panic, as is a sense of constriction (the suit) or suffocation (the gas mask). Crawling through a dark, enclosed space certainly isn't calming, either.

It was a bad ceiling, a difficult ceiling. While moving through a confusion of metal, dirt, and heat, I slithered over ducting, crawled across the rare ceiling support, and wriggled through the tiny spaces between girders and pipes. My tool belt snagged every few feet. I was pulling a heavy bundle of cable, the sweat ran into my eyes, and I could see my panted exhalations eddying a thick mist of fine dust in the beam of my flashlight. I spread out on top of some pipe and radioed my partner, James.

"James," I said, "I'm getting close to the drop. I've got about thirty feet to go, but it gets kind of tight up ahead, and I don't see another way to get there." The gas mask worked very well, filtering out objects as small as a bacterium. The masks also filter out all moisture. It was like breathing in the driest of desert air. I had to pause to work up some saliva.

"How're you doing up there?" James asked.

"OK. It's tight, though." My heart was racing. I was trying to keep it together. I just wanted to get it done and get out of there, get outside and breathe.

James was sympathetic. He knew. "Yeah, I looked up there earlier. Well, just take your time, move slow, and give me a yell when you get there."

"All right." I was breathing much too quickly. I tried breathing in through the mouth and out through the nose. I closed my eyes. It wasn't helping. I decided to rest on top of a wide sheet-metal duct. When I levered myself up on top of it, the duct began to collapse under my weight. As it dented it rang out with a hollow gonging sound. I could see the far wall where I would make a hole in the ceiling and pass the cable down to James below. As best I could tell, the only way forward was the top of the duct, which provided a straight line. But I could see that further down the space above the duct became a tortured tunnel of girder and pipe and wire.

It was the only way. It was my job. I started to crawl the length of the duct, which threatened to collapse under me. I didn't care if it did. In fact, I sort of wanted it to fall apart. Vacillating between fear and anger, I wanted to tear the place up or just fall through the ceiling, break an arm and collect disability until I could find a job fit for a human being. My suit caught and I shredded the leg by jerking it free. Finally, I came to a place where a girder crossed the duct, leaving a space of twelve or thirteen inches between me and my destination. I lay before it cursing to myself. Rather than do it on my stomach, with no way to pull myself through, I decided to go through on my back, reach over my head and grip the edge of the girder to slide my body across the sheet metal.

As I began to pull myself under the girder, bits of fiberglass disturbed by my shifting grip cascaded down onto my forehead and into my eyes. The sweat was running down the sides of my neck and suddenly, my mask caught on the underside of the girder and was pulled halfway off my face. Startled, I took a deep breath, caught a lungful of dust, and began to cough. Panicked, I let go of the girder and worked my hands into the few inches between my chest and the girder to try and reposition my mask before breathing any more of the dust. I began to hyperventilate once I got my mask back over my mouth. I was wedged in, my hands palm-down on my chest. Positioned like a corpse in a coffin, I suddenly became aware of the building as I had never been before; the terrible mass of the structure and the overwhelming number of rivets and bolts and steel ties and tons of concrete anchored a hundred feet below street level in the San Francisco hillside and the way that it would move in an earthquake, and I could see the rivets popping loose and the dust storm that would swirl in the space between the floors, and the girders twisting and fighting to tear themselves loose from each other, the walls opening and closing like the snapping jaws of a dying animal, and I knew that, in the position I was in, I wouldn't have a prayer.

I bit my lip hard to redirect my consciousness to something real, tried to control my breathing, and closed my eyes in denial that I was even there. I knew that I couldn't stay up there any longer or I would start thrashing around and put myself in a worse position. I wriggled out from under the girder and radioed James that I was coming out. When he asked what the problem was, I replied that I didn't know but it wasn't going to work and that I had to come out *now*. The tone in my voice left no room for discussion. James assented and again told me to be careful. There wasn't enough room to turn around in the space I was in so I flipped back onto my stomach and crawled out backwards, feeling the pounding of my heart in my bones the entire way.

After that day, the job turned into a struggle with fear and depression. Each morning, as the sun rose orange and winter pale and I sat in the 6:30 A.M. commute into San Francisco, I knew that I would spend a good chunk of my workday scared. The only upside was that I gave notice three weeks later, and that in those final weeks I began to cope with the fear that I felt whenever I put on a mask and climbed into the ceiling. Strangely, one of the most peaceful memories I have came out of those last few weeks. Once again I was in a tight spot, overheated and panicked, but not as badly as before, perhaps because I had an idea of what to expect. Rather than crawling for the hatch, I spotted a wider space about fifteen feet away, atop an air-conditioning duct, and made for that instead.

Once there I pulled myself on top of the duct, turned over onto my back and closed my eyes. Concentrating on the tingling chill radiating up into my body from the surface of the duct, I found that I could breathe slowly and easily. It was so quiet up there. I could hear a distant murmur of work being done in the room below me. But they were down there, caught up in their statistics and results, and I was stretched out and cool and comfortable and nobody but James, my coworker, knew where I was. For the first time, I turned off my flashlight and laid it down beside me.

As my eyes adjusted, I saw that it wasn't as pitch-black as I thought it would be, that the ceiling was imperfectly made and tiny shafts of light came through the pinprick holes in the ceiling tile, projecting a speckled radiance from below me and up onto the ceiling in my cramped but private space. The rays dappled the dusty surface above me like tiny stars in a night sky. And there, as I rested my body and the sweat cooled on my neck while considering this unexpected beauty, I fell into a deep and untroubled sleep.

*Geoffrey Alexander is a screenwriter and essayist who divides his time between northern and southern California.*

# With Enough Aspirin

## Living for Now in Pain's Company

LUANNE ARMSTRONG

Pain is a box, a space that shrinks and grows again during the day, a cage I carry with me, a tiger's cage with bars made of hesitation, effort, fear, exhaustion, a cage that keeps me holding on as I go down the stairs one at a time, that keeps me awake at night, twisting and turning to find a place to wedge my arms, hands, knees, in such a way that they will stop complaining and let me sleep.

Pain is a threat that makes me walk carefully. The dog banging my knuckles, toes knocked against the bedside, a too-outstretched arm, tiny mishaps in an ordinary day are sudden insights of pain. The thought of a slip, a stumble, a fall is unbearable. Every day, the dog and I make our way along the rocks by the beach: he flashing, fast, proud, me carefully behind. I will not fall, I think. I will not, cannot fall. I don't.

Pain is a new territory, a new country, a new culture with new customs. Now that I am privileged to enter here, I look for disabled bathrooms, sigh at hard chairs, shrink from any kind of journey. I discover that in getting on and off buses or airplanes, that long bottom step is a killer.

I count my other privileges: lying down in the long afternoons without guilt at work undone, windows or floors unwashed, wood uncarried, books unwritten. I count these: walking through the blazing autumn, trailing far behind the impatient dog, each step an achievement, each path a negotiation, coming home to peace, a long cup of tea, the chair piled with cushions, an evening full of silence.

Like any transition to a new culture, there has been a period of shock, of learning, of grief at the loss of what was once familiar, of reluctance to assimilate the new. People come at me using words once unfamiliar that now resonate with a power over my life: chronic fatigue, rheumatoid arthritis, immune deficiency, food sensitivities. These people mean well, but I sense they also wish to put me into another cage, one that fits, one that is comfortable, filled with the right kind of food and medication. But so far, none of them have fit properly. People shake their heads and sigh in dismissal.

Doctors also shake their heads. None of the tests are positive. My wonderful friend Carolyn, who is a doctor and alternative health practitioner, says "rheumatoid arthritis." She gives me books and articles and new pills to try. I read that rheumatoid arthritis causes exhaustion and depression as well as painful inflammation of muscles and joints. It's considered incurable. It's considered progressive. I shake my own head. None of this information feels like it applies to me. I don't feel like a person who is ill. I feel like a person in pain, acute pain that will soon go away, if only I can rest enough. But it doesn't, even when I try the pills and diets that people prescribe.

Today I have lunch with a new friend, someone with a bad back. He says he has been forced to learn all about pain management. I am immediately fascinated by the phrase and want to know what he means. From what he says, it seems mostly to have to do with lying down and taking lots of aspirin and not overdoing it on those days, in those moments, when the pain recedes and the sun comes out and life passes, for the closest it gets these days, to what used to be called normal.

He sits across the table. He holds his shoulders rigid. His lips thin and narrow when he moves. I know that face. I feel its mirror on my own. Some days the muscles in my face twitch. For a long while, when things were even more difficult than what passes now for usual, my face kept twitching into a kind of rictus. When I allowed it, followed it, looked in the mirror, it was like a snarl.

But I am a writer and this is a new place, so I try to follow my snarl to its conclusion, follow the new customs I am learning that are dictated by my body, which has turned foreign and tyrannical but stubbornly follows some wayward and vicious course of its own. I am taking notes. I am a new dweller here, passing through, I tell myself, on my way back home.

But as my familiar world recedes and shrinks and I move into this new place, people still come at me expecting the same person they have always known. They can't see that I have moved on and am living, for an unknown while, in a shadow country parallel to theirs. They peer in, see only the shadows shifting, offer, not unkindly, a litany of stories, remedies, miracles, suggestions, anecdotes, hopefulness. I go to lunch with others who are full of friendly advice and speculation. But advice and speculation blow away like air. They don't help me dress or clean my dreadful floor. I listen and thank them and pass on, silently cursing my own cynicism and what seems like everyone else's penchant for believing in miracle cures. Expensive miracle cures. If I believed, would they work? Is it only my lack of discipline, lack of faith, stubborn and stupid unwillingness to subject myself to regimes of fasting, cleansing, enemas, organic everything, exotic herbs that keep me from instant joy? I live in the country, grow and eat organic food, drink clean water that pours off the mountain. Can't I just endure this?

I struggle to find a stance that serves both me and my friends, try to find an easy place where they are comfortable and where I feel neither whiny, self-pitying, nor pitiable. At the same time, I struggle to find ways to say who I am now, to explain this new dimension in my life that takes my time and attention away from them and my life. I try to explain, but I am using a foreign language.

What they don't see is that not only have I moved into a new place, but I myself have been invaded. Pain is a kind of uninvited guest who has moved in and now refuses to leave, the rude kind who uses all the towels in the bathroom, eats all the food, makes life a hell of interruptions and never apologizes. But like anything else, one adapts, one gets used to it. I sigh and invite my guest yet again to dinner, knowing he'll stay anyway, that the length of his stay is neither up to me nor him, but to something more mysterious: my body, something called healing, a miracle.

Pain is something to which I learn to adjust. I allow for it, know what path I can walk, what steps I can take, what chairs I can sit in, what postures are almost comfortable. Some days, pain no longer feels just like itself, has a color, a texture, is distant enough to not make demands. It can persuade me what not to do, or we can make bargains. I always have the choice of pushing my way through the pain, whining busybody that it is, past it to get on with whatever it is I want and need to do. We negotiate. We live together, not on very good terms, but some uneasy, always-shifting middle ground.

The outer edges of this continuum are unknown. On one end is a mystery called death and on the other an equal mystery called health. My body has always served me well, now it cries out for rest, sleep, peace, stillness. I move through my world these days as an anomaly, a world I created when my body was mine to command, a farmer's world, a world of animals, gardens, hiking, friends, family, community, a busy world, full of demands needing time, energy, strength. At the moment, those are scarce, and my world shrinks.

I talk to pain, ask it who it is, what it wants, where it comes from, ask it for meaning. But pain is mute. People talk gently or uneasily to me about grief, pain, repression, about meditation and forgiveness. But pain is a puzzle not so easily teased into revelation. Like a Zen koan, it eludes understanding, focuses my attention on the moment at hand, on the next movement, the next chore, the next bits of time to be gotten through.

This focus is also a gift. With swollen and painful joints, no move is taken for granted. With a limited amount of energy, I can only manage a few relationships, a few projects. I am forced to concentrate.

I quietly divorce some of the more difficult elements of my large and noisy extended family. For now, I give up trying to save the world or even understand it. I'm happy enough to get through each day with enough accomplished to feel at ease when I lie down again each night. This may not look like much: the floor swept, dishes washed, the fire kept, a few letters and pages typed and stored, phone calls to friends. Enough.

Walking home through the wintry dusk, on my way to lie down and appease the noisy litany of aches and complaints from various body parts, I am thinking of ambition. For I am, as all writers are and must be, ambitious, for myself and my work. And this ambition, which has been a spur and an itch and a thorn and a variety of nasty little devils, has somehow receded and now takes a more balanced, dimly lit place in a corner somewhere.

This is new, but then, so much is. Astonished, I look at the sky. Well, I think, humbled at last, but no, not quite humble, but far more willing to accept with grace the gifts of each ordinary moment, to fumble my way through this life with less worry, less trepidation, less haggling with imaginary futures. I have wished for peace but never thought it would come as such a peculiar gift. How equally astonishing that since my body has developed such a variety of physical complaints, the black hole of emotional depression has gone.

I worry, of course, about my future and my writing and my tiny income. Occasionally I am acutely irritable and impatient. I am also aware of the lurking but unexamined fear that there might be no miracle, no healing, no transformation, that I might not again ride horses or take a bath or go hiking, go for long, leisurely swims, or even worse, be able to prune my sixty-year-old apple trees ever again.

I thought such activities were necessary to my existence. Now, I am surprised to see they are peripheral things, additions to my life that I once chose, but if necessary, can live without. I don't yet know what else I can live without. I don't want to know. But I know now the core of my life is lived on the inside, hidden, continues even when all such peripherals are stripped. The core of my life is not what I do, it is the harsh, white, twisted flame that burns and burns, gets me up each morning, makes me eat, stay alive. The core of my life is merciless, demands I keep living with as much effort and energy and skill as I can manage.

I accept all such gifts. I can't afford to be choosy. I accept my own humility, my hobbling walk, my painful lurching up from chairs, my grimace at others also bent and moving slowly through the supermarket, hanging onto the shopping cart. I accept them, perhaps not so much with grace, but at least with humor. If pain has made for me a cage, its walls are porous, full of light and color. And like all travelers, I know more than I knew. I have seen new facets of myself, and I have seen and learned some few small things about another country previously hidden and unknown to me. Now, I see and understand that it was all around me all the time.

Pain's largest gift to me, in return for its unscheduled stay, is that I have this lens to look through, this focus on each moment, this time when time itself is slowed, deliberate and sometimes, blessedly, gently empty of anything to do or think or be, when pain and I lie down together in the winter's dusk, with a book, a cup of tea, some aspirin, and long hours alone together and at relative peace.

*Luanne Armstrong is a writer and organic farmer on her family's homestead in British Columbia, Canada. She has published two novels, two books of poetry, and a children's book, as well as numerous magazine articles, short stories, and essays. She teaches creative writing at the College of the Rockies in Cranbrook, British Columbia.*

# What They Don't Tell You About Hurricanes

PHILIP GERARD

What they don't tell you about hurricanes is the uncertainty.

First it's *whether*. As in *Weather* Channel. There's been a rumor of a storm off the coast of Africa, and it's turned into a tropical depression. It churns across the Atlantic into the Caribbean and is upgraded to a tropical storm, winds at 40 or 50 knots, and the person in charge of such things gives it an androgynous name: Fran.

Will it hit us here on the southern coast of North Carolina? They can't tell. The experts. We've been through this before—Hugo, Felix, Marilyn, Edouard, Bertha. My wife, Kathleen, who grew up with California earthquakes, bridles at the lingering uncertainty, the waffling, a whole season of emergency. She wants it quick, bang, and over. But it doesn't happen that way. Hurricanes are big and slow and cyclone around offshore for a few thousand miles.

So the radar scope on the Weather Channel becomes familiar, part of the nightly ritual before going to bed, like taking out the dog and locking the front door. It becomes the first thing you do every morning, even before coffee. Watching the swirls of red and orange, a bright pinwheel of destruction. Checking the stats—wind speed, barometric pressure, latitude and longitude. We are at 34 degrees 12 minutes north latitude, 77 degrees 50 minutes west longitude. A degree of latitude equals 60 miles north or south. The arithmetic isn't hard.

Fran bangs into some islands from the vacation brochures and it's heading toward the U.S. mainland. But here in Wilmington, we just had Bertha, a direct hit. The eye sat over our backyard—you could look up and see the actual sky wound into a circular wall, like being down inside a black well, watching the stars out the top.

Surely, not twice in one season—what are the odds?

What they don't tell you is that hurricanes, like lightning, can strike exactly the same spot time and again. Fran is not the first storm. It's the second slam from a hurricane in eight weeks, and in the meantime it's rained torrentially almost every day. It's been a whole summer of violent storms, of lightning fires and local floods, of black-line squalls that knock down fleets of sailboats racing off the beach. The ground is so saturated we have had the lawn sprinkler system turned off all summer. Starved of oxygen, tree roots are rotting in the ground.

The longleaf pines that ring our property stand sixty and seventy feet high, two feet in diameter, precarious upright tons of wet wood, swaying already in the breeze. Their roots are soft in the spongy ground.

We've been set up. It feels like there's a bull's-eye painted on the map next to the words *Cape Fear.*

So it's *when.* Fran is moving at 14 knots, then 16, then wobbling slowly into a kind of hover. It's a monster storm, darkening the whole map of the Atlantic between Cape Fear and Bermuda, sucking up warm water and slinging it into windy horizontal rain. It's too big to miss us entirely.

It's Monday, the beginning of a long week. We fill up the bathtub, stockpile batteries and canned goods, locate flashlights and candles and matches, fill the truck with gas. Then we load all our important documents—passports, mortgage papers, insurance policies, marriage license—into a single attaché case and keep it handy. We take lots of cash out of the automatic teller.

Landfall of the eye expected Wednesday night, late. Wednesday is good for us, because Wednesday means south. Good for us, bad for Charleston. Hugo country.

We wish it on them. Me, Kathleen, the neighbors who drift back and forth between houses just to talk out loud, just to look at the sky. We feel bad about it, but we wish it on them anyway. If we had real magic, we would make it happen to them, not to us.

But Fran wanders north, following Bertha's path, and on TV they change the *when:* Thursday night, after midnight. Because our Beneteau sloop, *Savoir-Faire,* is moored in a tidal harbor, we pay attention to the tides. Low tide will be at 9:34 P.M. From then on, the tide will rise one foot every two hours until 3:29 A.M. By mid-afternoon all of us whose boats remain in the community harbor at the end of our street are lashing on extra fenders, strapping lines to the pilings, watching the water lap at the bulkhead separating the marsh from the harbor.

I'd take the boat out of there, drive her to safety, but where? It would take eight hours to get down the waterway and up the Cape Fear River, and I don't know the hurricane holes there. I'd be stuck on the boat, away from my wife, in the low-country wilderness with a 3- to 5-knot current pushing dangerous debris down the river at me all night long.

*Full-force Fran aims for coast,* says the local newspaper front-page headline.

Everybody is thinking the same thing: *Don't let it come ashore at high tide.*

We speculate nervously about how much the tidal surge will actually be in this protected harbor, blocked from the ocean by a large, developed barrier island—Wrightsville Beach—a channel, a spoil island, the Intracoastal Waterway, and finally a hundred yards of marsh that are dry land at low tide.

Nobody knows.

Our docks are the floating kind—they can float up on their pilings another nine feet, and all will be well. All of our boats made it through Bertha without a scratch—85-knot winds and a tidal surge of six feet.

There's the standard hurricane drill: strip all sails, remove all windage-making gear—horseshoe buoy, man-overboard pole, lifesling. We all help one another. Nobody has to ask. While unbending the large full-battened mainsail, I bang my new racing watch on the boom gooseneck and break it. A bad portent.

We retreat across the causeway to our homes, where the power has already gone off, as the rain becomes torrential and the wind begins to blow in great twists of energy. It has started. So we have an answer to *when*. An hour later, when Fran comes howling down on us out of the ocean, it's *how hard*. As we huddle indoors and listen to the roaring, the question becomes *how long*.

*When, How Hard, How Long:* the trigonometry of catastrophe.

The answer is 8:05 P.M., almost dead low tide.

The answer is sixteen feet of surging water anyway and winds of 105 knots.

The answer is fifteen hours.

Some of the clichés turn out to be true.

The rain really is *torrential*, as in *torrents*.

A hurricane *does* sound like a freight train. Exactly like. If you were lying between the rails and it went roaring along over your head all night long. It really does *roar*. Like whatever is holding the world together is coming apart, tonight, this minute, right here, and you're smack in the middle of the program.

And your mouth really does go so dry with fear you can hardly talk.

The great trees cracking and tumbling to the ground in the roaring darkness really do sound like an artillery barrage—*crack! crack! whump! whump!* It takes italics, exclamation points, boldface clichés to tell about it. The house shudders again and again. Our house has too many large windows, so we run next door to wait out Fran with our neighbors. We're sitting up with them in their living room drinking any liquor we can get our hands on—vodka, beer, wine, rum—and each shudder brings a sharp intake of breath, a little cry. You can't help it. You laugh and make jokes, but it feels bad, and the feeling gets worse every minute. The kerosene lanterns don't help. They make Halloween light. Eerie, spooky light.

There are times when you have to dodge out into the maelstrom of wind and flying debris and back across the lawns to check the outside of your house, to clear the storm drain and prevent flooding of the lower story. It's stupid, especially in the pitch blackness, but it feels like something you have to do. The world is way out of control, but you're still responsible.

There are freaky contradictions of nature. Paradoxes of chance. A massive oak tree that has weathered three hundred years of storms is ripped apart by the wind, literally twisted out of the earth by the roots. The next lot over, a pair of forgotten work gloves left to dry on the spikes of a picket fence are still there in the morning, and so is the fence. Dry.

The wind blows strips of new caulking out from between the casement windows but leaves intact the plastic tarps you nailed over the open sides of the upstairs porch.

There are amazing feats of heroism and survival. A man on one of the beach islands sends his wife and kids to the shelter, remains behind with their dog to finish boarding up the house, then the only road off the island overwashes, and he's cut off. He grabs his dog in his arms and ropes himself to the house, and all night long he and the dog are bashed against the house by water and wind, but they make it through. The dog was a boxer.

Lightning strikes the home of an old couple and it catches fire. Two young men appear out of the storm, attack the fire with a garden hose and keep it from taking the house until the fire trucks arrive, then disappear. Nobody knows who they are or where they came from. The old couple believes they are angels.

There are tales of death. A man is seen stepping onto his front porch as the hurricane hits. They find him in the morning miles away, floating face-down in the Intracoastal Waterway. A woman rescued from a mattress floating in the marsh dies anyway.

For a week afterward, urban rescue workers prowl the wrecked homes along the beach with dogs, sniffing out the bodies of the ones who wouldn't leave.

It's also true, the cliché about the capriciousness of nature and about blind luck. Three marines in a Mustang are swept off the road by the rushing water. One is washed to the far shore and stumbles into a shelter. The second clings to a tree limb for nine hours until he is rescued. The third drowns.

There are things that are outrageously unfair. A family down the street gets flooded out on the ground floor. They scramble upstairs ahead of the surge. But the battery of their brand-new car shorts out in the rising water, and it catches fire. The garage underneath the house burns. Soon the whole house is burning. Incredibly, at the height of the hurricane, the volunteer firemen arrive. They maneuver their pumper through waist-deep water. But they can't get the electric garage door open and have to axe it down. And by then the family is smoked out, the house is partly destroyed, the car is a hulk.

Hurricane, flood, fire, all at once.

Thunder and lightning come in ahead of the hurricane. Tornadoes spin off the leading edge like missiles, knocking out bridges, tearing holes in houses, twisting trees out of the earth and flinging them into power lines.

Biblical stuff.

*Furious Fran Unforgiving,* the local newspaper says, again on page one, unable to let go of the corny habit of alliteration.

What they don't tell you about hurricanes is the heat.

The oppressive stillness of the stalled atmosphere the day before the winds start. The hundred degrees of swampy humidity the day after, before the torrential rains resume. The air-conditioning is off, the windows are latched down tight. An hour into the storm, everything you touch is greasy. You put on a fresh shirt and sweat it through before you can fasten the buttons.

And then the bees arrive. Swarming, disoriented, stinging bees gone haywire. Bumblebees, wasps, yellow jackets, hornets. I'm no entomologist—they all sting.

After a hurricane, the radio warns, that's when the injuries start. Bee stings are number one, followed by poisonous snakebites and chainsaw cuts.

When the rains resume a day and a half after Fran passes, the yard is jumping with frogs and toads. Little bright-green tree frogs with suction cups on their feet, smaller than a penny. Black toads the size of your fist. Giant croaking bullfrogs that splash around like rocks. Rat snakes. Water moccasins. Copperheads.

What's next—locusts? Well, not exactly: crickets. By the millions. All over the debris, the backyard deck, the wrecked boats.

But the birds are gone.

The water is off.

After sweltering hours clearing the tree limbs out of the road, pulling limbs off cars and shrubs, dragging downed trees off the driveway, raking the mess off the steps and walks and deck, my wife and I shower by pouring buckets of cold water, saved in the bathtub, over our soapy heads and bodies. We are scraped and cut and bruised and stained with pine resin that does not wash off. Every pair of shoes we own is wet and muddy and will not dry. The house is tracked with mud and debris, and a lethargic depression sets in—part physical exhaustion from relentless manual labor in the heat from two sleepless nights in a row, part emotional exhaustion. Grief.

We were luckier than many. It just doesn't feel that way.

When the power comes back on, it's like a religious experience. Everything becomes possible again—bright lights, cool air, television news, ice.

Then after a few hours it goes off again.

What they don't tell you about hurricanes is that the Big Hit is the beginning, not the end. Fran has swept on up the coast, taking the Weather Channel and CNN with it. On the networks, things are happening in Bosnia, Chechnya, Indonesia.

Here in Hurricanelandia, it's raining eight inches in three hours on top of ten inches that came in with Fran. They predict it will rain for another week. All the low-country rivers are cresting, shouldering through the wreckage of human cities toward the sea.

Our house is an island surrounded by rushing water two feet deep, and it's back out into the storm wearing Red Ball boots, clearing out clogged gutters on an aluminum ladder, counting the seconds between lightning and thunder, counting how long to dare such foolishness. Then slogging out onto the muddy access road behind the house to rake out the clogged storm culvert, trying not to get carried into the muddy water.

On the local radio, the jocks are chatting about this and that and the other, but for hours nobody gives a weather report. When will it stop raining? *Will* it stop raining? The phone is working. A friend from across town, where they have power, calls. Look out your window—is it raining there? The edge of the cloud is moving over us now, she says, and there's sun behind it.

The water recedes, and now it's time to clean out the flooded garage. At dusk, the generators go on. It gets dark and noisy. We will wake to the lumber camp sound of chainsaws.

For weeks and weeks.

What they don't tell you about a hurricane is that it just seems to go on and on.

But the worst of it is not captured on the awesome helicopter videotape of destruction. The worst of it is waking up to the new stillness of the morning after, when the wind has finally quit and the rain has slacked and the sun may or may not be up yet, the sky is just a gray slate of clouds.

Overnight, the world has changed in some important, irrevocable way. You can just feel it.

My neighbor John is standing outside waiting. "You ready?" he says, and I nod.

Half a mile away, the approach to the harbor is littered with dockboxes, paddles, small boats, life jackets. Like a shipwreck has happened to the whole neighborhood. The houses by the harbor have taken a beating. A forty-four-foot sportfishing boat lies on its side on a front lawn, and my stomach turns. That's how high the water rose.

A few nights earlier, I had stood on our dock talking quietly with an old friend, admiring the sleek, trim lines of *Savoir-Faire* under starlight, feeling lucky. Thirty-two feet of beautiful racing yacht, a dream of fifteen years of saving come true. I'd take *Savoir-Faire* out onto the broad back of the Atlantic and race her hard, rail down or just jog along in mild breezes, clearing my head, sharing her with friends or filling up with the good strength that comes from working a yare boat alone.

The harbor was demolished. Boats and docks were piled up like a train wreck. Boats were crushed, sunk, broken, smashed, aground. Some were simply gone.

Out in the middle of the harbor, alone, *Savoir-Faire* lay impaled on a piling, sunk by the bows, only her mast and transom rising above the dirty water.

*Philip Gerard is the author of three novels and two books of nonfiction, including* Creative Nonfiction: Researching and Crafting Stories of Real Life *from Story Press. He lives with his wife Kathleen Johnson in Wilmington, North Carolina, where he directs the MFA program at the University of North Carolina, Wilmington. This past June, he sailed his new boat,* Suspense, *from Miami home to Wilmington.*

# The Lightning in My Eyes

JEAN HANSON

We're driving through South Dakota when I see the tall grass on the side of the road turn liquid. Then the plains come alive: They breathe and relax, breathe and relax. We could be in a boat on a golden ocean for all the dipping and swaying. After a while, the horizon flickers and sends up a filmy light. The air itself is viscous, moved by wind, distorting the landscape.

I close my eyes. It's been hot, an uncomfortable day for a long drive. We're going to visit my grandparents in the tiny town of Wilmot. I know this, of course, and yet at moments I feel I've left it behind and I'm observing myself, Jean, as if she is some curious artifact. When I open my eyes, flashes of lightning bolt across the sunny road. My husband Chris doesn't slow the car. The lightning is in my eyes, not in the atmosphere.

My grandmother, a woman with a face wrinkled like a dried peach, walks down the painted cement steps to greet us. We sit in the kitchen, and when I talk, my voice comes from the other side of the room, as if I am a ventriloquist. I press my index finger to my thumb, but my digits move through each other like gelatin, tingling. I am of the world, but I'm not.

My grandfather is thrilled with Chris, whom he keeps calling Pete.

He displays the wooden coat hangers he carves, the best hangers in the world; he shows off his immaculate Buick, the best car in the world; he serves us Riunite, the best red wine—no, in honor of Pete, my husband of Sicilian heritage—the best *Italian* red wine in the world. I hear this as pleasing but insignificant background static.

Dinner is circular. Voices: first Jean's, then Chris's, Grandma's, Grandpa's, Jean, Chris. Serving dishes passed around and around the table. Forks moving in slow rotations around our plates. These spherical rituals are an orchestral accompaniment, my consciousness the melodic line moving above it. I am disengaged and hovering, monitoring Jean and the others. Chris puts his hand on my arm, as if to ground my flight, but I resist: I'm rising. My head is helium. I'm taking everything in, seeing patterns and meaning. I'm on the verge of understanding the whole, crazy, profound lot of it.

Only later is there nausea. In the guest room, I crawl into the bed my father was born in. He has died two months before. His high school graduation picture is on the bureau, and he watches as I leave his realm, the spirit world, and move into pain as pure as frozen winter, an icicle poked in my forehead.

If I ask you to define *migraine,* you will call it an excruciating headache. Well, yes. And no. For those like me—among the distinct minority who suffer "classic" (with aura) rather than "common" migraine—the journey is more circuitous.

On days when you're singing through the mundane details of life, admiring the warmest chambers of your husband's heart and feeling lucky, you may be on the verge of a migraine. The migraine precursor is often a nearly euphoric sense of well-being—George Eliot described it as feeling "dangerously well." It can also manifest as apprehension, a texture of strangeness. You can't shake the notion that the world is being dismantled, its edges unraveling.

Sometimes your husband knows before you do. He's noted a certain posture in your sleep and a slowness in your reasoning. Your sister hears it in your legato voice: There's no melody, she says, you've gone flat. Then a glass slides from your hand. You mail your wallet.

Warning. If you're in the supermarket, abandon the shopping cart. Drive directly home. Narrate out loud: green means go. Red means stop.

Your body is a miser now, retaining fluids, keeping all to itself. Your face turns pale as milk, and half-moon circles, blue like bruises, appear under your eyes.

Next, you experience "aura," a complex neurologic mischief. The brain has a fine time of it—entertaining you or terrifying you, depending on your disposition. With the right attitude, you can traipse along, admiring the chicanery of your cortex. Consider *Alice Through the Looking Glass,* for instance, a fantasy based on Lewis Carroll's migrainous visual disturbances. Carroll perceived distortions of size: Lilliputian (diminution), Brobdignagian (enlargement), and zoom vision.

The religious visionary Hildegarde didn't see "The Fall of the Angels" or "The Living Light." "The Aedification of the City of God?" Hardly. Yet these visions were real; so was the lightning in my eyes. The scotoma of migraine is measurable as swarms of phosphenes cross the cortical field.

During aura, you might, as I once did, get lost in a building where you've worked for a year. You wander the halls. Just where is that office of yours? It takes an hour to find. Then you close and lock the door, turn off the lights, and ring your husband, relieved to have mastered the complex trigonometrics of dialing a phone. The husband who answers is yours, though you can't quite place his first name.

You request the television be turned off with the words "Petal shower ringing."

Your vision narrows, as though you are viewing things through the wrong end of binoculars.

You don't know the year, your address, who the president is.

You look in the mirror and are shocked: your eye has moved. On closer inspection, your whole face has been segmented and rearranged. You're a living portrait conceived by a cubist painter.

Only after your brain shows off, establishing who is in charge, do you move to the next stage, with its nausea and vomiting; sensitivity to light, sound, and smell, and the legendary headache. To imagine the severity of the pain, consider the age-old remedies for it. They purge you, bleed you, lobotomize you, chop a chunk from an artery, drill a hole in your scalp. They loop a hangman's noose around your skull or anoint you with moss from a statue's head. They bind to your brow a clay crocodile stuffed with magic herbs.

And none of it helps.

Migraine resolves when your body becomes generous once more. You urinate copiously, then receive a gift of sleep. Some migraineurs are exhausted after an attack. Many, like me (also Freud, who credited his good health to "the regulatory effects of a slight migraine on Sundays"), are renewed. We appreciate. We see clearly. We get a lot done.

While writing this, I am downed. My husband and I are supposed to make a weekend foray, but we defer to my defective neurons instead. A different journey begins. Chris comes quietly into the room, offering crackers and Coca-Cola.

The weekend drones on and on. "And screen'd in shades from day's detested glare," says Pope, "She sighs forever on her pensive bed, Pain at her side, and megrim at her head."

I am not only sick, I am guilty. This is my fault, I know. Haven't the pundits—from Pliny the Elder to Aretaeus the Cappadocian, all those Romans and Greeks, the pitiless Victorians—told me it is? This "mygrame and other euyll passyons of the head" is due to my bilious humours, my hereditary taint, a hysteria of my uterus. My nervestorm is caused by masturbation, violent passions, and errors of diet. Experts of the 1930s note my retarded emotional makeup: I am perfectionistic, inflexible, and obsessive. By the late fifties, though my sins of ambition and rigidity are set in stone, I'm no longer bereft of charm. The oft-quoted Alvarez describes the small trim body, firm breasts, stylish dress, quick movements, luxuriant hair, and eager mind of the female migraineur. "These women age well," he insists.

There is comfort in the conviction that you can cure your disease by refashioning your personality. If I simply become more easygoing, I think, if I don't push so hard or stay up so late . . . but today's research dashes these hopes.

Migraine strikes the indolent as often as the driven, the sloppy along with the neat, the profligate with the parsimonious. It's an organic dysfunction, and like all biochemical disorders, it's poorly understood. Essentially, migraineurs don't have effective brain filters. Our volume controls are turned up, and we let in static. This makes the hypothalamus get kooky, alarming the reptilian brain and flustering the limbic system. Circuits break, sparks fly, and neurotransmitters run amok. The poets of medicine call it the chain reaction, the cascade, the avalanche activated when we encounter any number of everyday "triggers."

Say, for instance, that you're driving at night in a snowstorm. Your headlights illuminate swirling flakes, which seem to give birth to a million fireflies. You step on the gas and these icy stars part like a fast-forward film of the big bang theory. Bang. Big bang. This is a visual trigger that can bring on a migraine.

Or you're caught in the California Santa Anas. You face the Argentine zonkas or the Swiss foehns. These dangerous hot winds are migraine triggers all. (Go to France or Canada, instead. The cool mistrals and chinooks have no effect.)

Say you forget to eat. Or you drink red wine.

And remember, someone once told me, to avoid the three c's: cheese, chocolate, and citrus. Of course, spurn the vile potato and the wicked garbanzo bean. Consume no bits of bacon, bites of hot dog, or slices of pumpkin pie.

Also thought to bring on migraine: too much sleep, too little sleep, lights, glare, altitude, stress, sex, garlic, smells, noises, humidity, travel. . . .

You may agree to give up Chianti and sauerkraut, but not vermicelli alla puttanesca. Not Argentina. Not sex. The world causes migraines, and who wants to avoid the world?

So you outwit it. The pharmaceuticals you use are beta *blockers,* calcium channel *blockers, anti*serotonins, *anti*histimines, *anti*convulsants, *anti*depressants. The self-help books are entitled *Overcoming . . . Fighting . . . Beating . . . Victory Over Migraine*—as though, in order to live, you must do battle with your own brain.

But still the hot neural storm disrupts your life, like a twister uprooting a tree—maybe not four times a week, but twice a month. In moderation, you can accept the brain's imperialism. You can stand back and watch your mind build its cathedrals.

Six or seven years old, I am sitting on the sidewalk with my cousin Sharry. The steps are cracking, and pieces of chipped cement sparkle like little gems. Then, a shift, as if a cloud has covered the sun. I look at the large, tall trees spread evenly across the lawn. I want to name them, bring to mind the sound that will tell their shape, their smell, their distinctive color. But the words, which I knew minutes ago, are no longer available. A shiver. A veil descends. I am trapped in a small, thin, mosquito-bitten body. Whose? I don't know. I've lost acquaintance with myself, but gained an opening to the unknown. I'm curious about the disposition of my soul, though I'm too young to put it that way. "Who *am* I," I say to my cousin without a question mark. I mean so much. But she laughs, expecting a foolish childhood game.

I was experiencing *jamais vu,* a sense of unfamiliarity, of a newly made world, a phenomenon, like *deja vu,* common in migraine aura. The incident so affected me that I wrote about it repeatedly as a child, until years later I found out it was simply a neurologic trick. But by then it was too late. The episode had been infused with meaning. Dutiful student of my cortex, I wondered, what am I to make of this? I learned that reality was not stable: it was subject to improvisation. I learned I could be exiled from my own life. I learned I was different from my cousin Sharry—isolated, a drifting spirit. So where did the pathology of migraine end and the development of personality begin?

Oliver Sacks, writer and doctor, calls migraine aura a veritable "encyclopedia of neurology," with its aphasias, paresthesias, paralysis, odor hallucinations, and amnesias—but it is much more. Aura delivers a whispered message of deep personal significance, more eloquent than language, more urgent than reality.

It's summer in North Carolina. Chris, Richard, and I are driving in the mountains. As we pass a town bathed in sun, sunk in a valley, I begin to feel its rhythms, its nine to five, its peculiar, humming pulse, its wooden floors grooved with a hundred years of use. It's as if its collective memory has been installed in me.

At a stop, I look at a billboard. Something is awry, undone. I can't read. I try to sound out words, concentrating like a first grader for whom each letter is a new challenge. But if I look at one part of a word, another disappears. Soon, I see the problem: there's a hole in the world and whatever I observe falls into it. My very gaze is fatal.

I touch my cheek and lips, rub my fingers together, and brush my hand down my arm. A weather front is spreading through my body, numbing one side of it. For a moment, I wonder if I've had a stroke, but dismiss this as too theatric. And though I'm not sure, I barely recall—haven't I taken similar journeys of mind before?

At a stream, Richard and Chris disappear with fly-fishing gear. I move toward the sound of water. The vegetation is lush, and though there's nowhere to sit on the bank, I see a huge boulder in the stream. I long for this rock and wade out unsteadily, conscious of a dizzying sweep of water.

I climb onto granite: solid, steady, old as the continent. It's been a long week. Chris and I have been sharing a beach house with our old friends, Rich and Carol, who are smoking again. Their children have become restless toddlers. The oldest holds her little sister's hand in a door and slams it shut; the youngest grabs shrimp off our dinner plates. It's been hot, and I haven't slept well.

But here, the air is cool, and this moment a ballet—me pirouetting on a rock, the water twirling and upstream, Richard and Chris on the water, casting, happy to be together. Suddenly it's quite clear: I've seen this exact dance before. Perhaps I've even rehearsed it. All our steps have been precisely and lovingly choreographed. It's a work of art, orderly and resolute. This is its performance.

Now I get close to the stone, feel its warmth on my cheek as I stretch across it. Soon, the pain will come, with its paralyzing but cleansing purity, and then sleep. When I awake, the sun will be in a different place in the sky. I'll be grateful for good friends and their children. My eyesight will be renewed; the edges of my life newly distinct. And though the potent elixir of knowing more than I am meant to will have disappeared, the memory of it will linger. The world will be magical and mysterious. After all, I've been its creator today: I've visited a variation of it—remade by my own mind.

***Jean Hanson** has published work in a variety of literary magazines including, most recently, a short story in* Puerto del Sol. *A graduate of the Iowa Writers' Workshop, she has received an Artist Fellowship from the Colorado Council on the Arts, the Hackney Prize in the Short Story, and a Writer's Exchange Award for emerging writers from Poets and Writers, Inc.*

# Chimera

GERALD N. CALLAHAN

Last Thursday, one of those gray, fall days when the starlings gather up and string between the elms around here, my children's mother—dead ten years—walked into a pastry shop where I was buttering a croissant. She ignored me, which she always does, ordered a plain bagel and an almond latte, picked up her food and, without a glance at me, walked out. The starlings chittered, the day frowned, and I went back to buttering my croissant.

Just after her suicide, I saw this woman often—in towns where she never lived, walking her Airedales in the park, eating poached eggs at Joe's Cafe, sweeping grass clippings from her walk on Myrtle Street, stepping off the Sixteenth Street bus. We get together less often now. But when we do, like this morning, her image is as vivid as it ever was—her dark eyes as bright, her odd smile just as annoying.

I'm not crazy.

I know it isn't her, this woman I see. After all, she's dead, and I myself gave her ashes to my son. So it is another, a stranger, transformed by some old film still flickering through the projector inside my head. I know that. But every time I see her, it takes all that I have to stay in my chair or my car, to hold onto myself and not run after her calling out her name.

Some of this I understand. When something or someone is suddenly stripped from us, it seems only natural that our minds would try to compensate. Minds do that. If they didn't, we might be sucked into the vortex ourselves. That part, I grasp. I'd have thought, though, that in a year or two, the films in my mind would fade and break, and the tear in my life would scar and close like any other wound. And I expected, as the fissure closed, that my first wife would disappear.

I was wrong.

All the pieces of human bodies fit (more or less) into eleven systems—endocrine, musculo-skeletal, cardiovascular, hematologic, pulmonary, urinary, reproductive, gastrointestinal, integumentary, nervous, and immune. So there are a limited number of places where someone could hide something inside a human body. And so far as we know, only two of the body's systems, immune and nervous, store memories—fourth birthdays or former wives. That narrows it even further.

Most of us don't for a moment associate immune systems with hopes and fears, emotions and recollections; we don't imagine that anything other than lymph—the pale liquid gathered from the blood—is stored inside of thymuses, spleens, and lymph nodes. The business of immune systems is, after all, not hope, but immunity—protection against things like measles, mumps, whooping cough, typhus, cholera, plague, African green monkey virus, you name it.

But immune systems *do* remember things, intricate things that the rest of the body has forgotten. And the memories stored inside our immune systems can come back, like my first wife, at unexpected moments, with sometimes startling consequences.

My grandmother had a penchant for saving things. She had grown up in a very poor family and believed nothing should be wasted. On the plywood shelves of her closets, Mason jars that once held apple butter or pickled tomatoes were filled with buttons, snaps, paper clips and strips of cloth, seashells, rubber bands, pebbles, bobby pins, and cheap, shiny buckles—everything she'd ever come across that she thought might be useful someday.

Immune systems do that, too—believe that most everything they come across will be useful again someday. Grandmother used Mason jars, immune systems use lymph nodes. Immune systems collect bacteria, parasites and fungi, proteins, fats, sugars, and viruses—the stuff that falls through the cracks in our skin. Human skin is like nothing else in this universe. It tastes of sea salt and the iron inside of men and women. Its touch arouses us. Skin is cream, sand, teak, smoke, and stone. But mostly, skin is what keeps us apart from everything else on this planet, especially everything that might infect, infest, pollute, putrefy, and possess us. First and foremost, it is our skin that allows us to be here as individual men and women in a hungry world. Skin keeps things out—things that would eat us for lunch. And skin keeps things in—things we couldn't live without.

But skin can break down, get punctured by knives and needles or scraped off by tree limbs and tarmac. When that happens, we'd die without our immune systems—abruptly. Immune systems deal with the things that crawl through the holes in our skin. They label the intruders as dangerous, round them up, and destroy them. And immune systems never forget the things they've seen beneath our skin because they believe that one day those things will be back.

That's how we get to be adults—immunological memory. That's also how vaccines work. Until a few years ago, children in this country were regularly injected with cowpox, also known as vaccinia virus. Vaccinia virus is very similar to the virus that causes smallpox, with one important exception. Vaccinia virus doesn't cause the disfigurement, illness, and often death caused by smallpox. But as Edward Jenner discovered in the 1700s, people (in Jenner's case, milkmaids) who have been infected with cowpox don't get smallpox. A miracle. Immunity to cowpox protects a child from smallpox. That's because, even though their personalities are very different, smallpox virus and vaccinia virus have a lot of physical features in common. Immune systems that have learned to recognize and destroy cowpox virus also recognize and destroy the look-alike smallpox virus before it can do harm.

And immune systems remember. They remember each and every miracle, and remember them for a lifetime. A child vaccinated against smallpox virus will make a much more rapid and specific response on a second encounter with that virus than will an unvaccinated child. And the rapidity and specificity of that second response is what saves the vaccinated child's life.

Immunological memory is a simple memory of a tiny virus, but a memory powerful enough to have ended the devastating disease of smallpox on this planet. In essence it is no different from the memory that pulls our hand from the flame a little faster the second time, the memory that guides the cleaver beyond the scars on our knuckles or the memory of a first love lost.

The way immune systems do this is extraordinary. Lymph nodes are little filtering stations strung throughout the human body. Lymph nodes monitor the fluids of the body—mainly lymph and plasma—for infections. When something out of the ordinary is detected, it is usually the lymph nodes that remember and initiate an immune response.

Every time we are infected, a little of the bacteria or virus that infected us is saved in the lymph node where it first arrived. By the time we're adults, lymph nodes are filled with a bit of most everything we've ever been infected by; our lymph nodes are the repositories of our infectious histories. Just like my grandmother's jars, our immune systems sort this growing mass of memorabilia and remind themselves of what they've seen before, what they are likely to see again, and what they mustn't forget.

Mustn't forget, but mustn't hold too close to the surface, either.

Because, just like some of the memories lurking in our brains, an inappropriate recollection can hurt or blind us, sometimes even kill us. Those things we suppress.

Some viruses and bacteria stored inside our bodies are intact and alive. The only thing keeping us from having the same diseases all over again is the constant vigilance of our immune systems. Through that vigilance, all of those things hanging around inside us are kept in check, are suppressed to the point where they can help us remember, but they cannot cause disease. Memory with a mission, selective recollection and suppression.

Lots of things can distract immune systems, though—drugs, malnutrition, stress, age, infection. When these things happen, immune systems can forget for a moment all those deadly things packed away inside of us. Then like minds in panic, immune systems can become confused, forget which memory to recall, which memory to suppress, and the past can flare inside of us. When that happens, our very survival depends on our ability to regain our balance, to enhance some recollections and suppress others. A particularly pernicious example of this is shingles—a severe chicken-pox-like rash that usually appears across the ribs beneath the arms, but may also grow in the eyes and lungs. It is most commonly a disease of the elderly.

People can't get shingles if they weren't infected with chickenpox, usually as children. Shingles and chicken-pox are caused by the same virus—varicella zoster virus. When we get chicken-pox, our immune systems and (interestingly) our nervous systems store a few leftover varicella zoster viruses for future reference. Later, when age or illness or depression distracts our immune systems, the virus begins to multiply again. Then the virus may blind us, may even kill us. This is shingles—a blazing memory of chicken-pox, a childhood disease—a thing we wish we could forget.

So immune systems, like minds, are filled with memories—vivid, painful, sometimes fatal memories. The fragments of a life lived, bits and pieces of the past. And sometimes immune systems lose control of this smoldering wreckage, and old flames flare anew.

Within me, then, is there a woman living in this ruin, a woman who walks and speaks exactly like my first wife? It is, of course, impossible to answer that question. No one understands nearly enough about wives and immune systems. But it isn't, as it might seem, an entirely stupid question. Among the things we regularly trade with our wives (and the rest of our families, for that matter) are viruses—colds, flus, cold sores, to mention only a few.

Enveloped viruses—like those that cause flu, cold sores, and AIDS—are so-called because they carry with them an "envelope" of lipids and proteins taken from the host cell (the cell they grew up inside of). And many viruses also carry within them a little of the host cell's nucleic acids—DNA or RNA—the stuff of genes. Some of that DNA or DNA made from that RNA clearly gets incorporated into our chromosomes and begins to work inside of us. That means that each time we are infected with one of these viruses, we also acquire a little of the person who infected us, a little piece of someone else. Infection as communication. Infection as chimerization. Infection as memorization.

Perhaps that seems trivial—a bit of envelope here, a little DNA there. But over the course of an intimate relationship, we collect a lot of pieces of someone else. And a little of each of those pieces is stored in our lymph nodes and in our chromosomes. Until. Until the person we've been communicating with is gone, and we stop gathering bits of someone we love. For a few days or weeks, everything seems pretty much like it was. Then one day, a day when for no apparent reason, our defenses slip just a little, and a ghost walks through the door and orders an almond latte.

Nervous systems don't appear to store memories in the same way immune systems do. Most neurologists and neurochemists believe that memory within the nervous system involves something called long-term potentiation or LTP—a means by which certain nerve pathways become preferred. Because of LTP a particular trigger—a picture of Aunt Helen—becomes likely to stimulate the same nerve circuit—the smell of cheap perfume—every time. But in general, how nervous systems store and recall memories isn't very well understood.

Human memory has been divided into two broad categories—declarative memory (explicit, consciously accessible memory: What was the name of the cereal I had for breakfast?) and emotional memory (often subconscious and inaccessible: Why was I so frightened by that harmless snake I saw today?). But there is evidence for a third kind of memory as well, something I'll call phantom memory, memories that come from some place beyond or beneath declarative and emotional circuits.

I'm pretty confident that declarative memory had nothing to do with my first wife walking in on me as I was buttering my croissant last Thursday. I'm less certain about emotional memory. And I am deeply intrigued by phantom memory.

People who have had arms and legs removed often experience phantom limbs—a sensation that the arm or leg is still there, sometimes a very painful sensation. This feeling is so real that people with phantom hands may try to pick up a coffee cup just as you or I would. People with phantom legs may try to stand before their declarative minds remind them they have no legs. The missing limbs seem completely real to these people and as much a part of themselves as any surviving appendage—even when the phantom limb is a foot felt to be dangling somewhere below the knee with no leg, real or phantom, between the ankle and a mid-thigh stump.

Some of those who have studied phantom-limb sensations argue that these are only recollections of sensations "remembered" from the days before amputation. But children *born* without limbs—children who've never experienced the sensations of a normal limb—experience phantom limbs. Clearly, these phantoms are not simple recollections of better days. Instead, the presence of phantom limbs in these children suggests that some sort of prenatal image—some template of what a human should look like—is formed inside our fetal minds before our arms and legs develop, before even our nervous systems are fully formed. If at birth our bodies don't fit this template, our minds or brains attempt to remake reality, twist it until it fits what our minds say it ought to be.

No one knows where phantom memories reside. Often, phantom limbs are exceedingly painful, so physicians have tried to locate the source of the sensations and eliminate them. Spinal chords have been severed, nerve fibers cut, portions of the brain have been removed. Some of these, sometimes, caused the pain to disappear, but it usually returned within a few months or years. And none of these treatments routinely caused phantom limbs to disappear.

Occasionally over time, phantom limbs will disappear on their own, though almost never permanently. The limbs usually return—in a month or a year or a decade. And when they do, they are just as real as the day they first appeared, or disappeared.

Phantom memories aren't always memories of limbs either. People who've lost their sight describe phantom visions: not recollections, but detailed images of sights they've never seen—buildings, burials, forests, flowers. Similarly, some people who've lost their hearing, Beethoven being one, are haunted by complex symphonies blaring in their ears.

No one knows how much of our reality comes to us from the physical world and how much "reality" we create inside our own minds. If we were to analyze, using something like a PET scanner, all of the nervous activity occurring at any given moment inside a human body, no more than a fraction of a percent of this activity would be directly due to input from the senses. That is, only a tiny portion of what our nervous systems are occupied with, and by inference only a tiny portion of our thoughts, are direct results of what we see, hear, taste, smell, or touch. The rest of it, the remainder of our mental imaging, begins and ends inside of us. How that affects our "reality" isn't clear.

But it is clear that much of what originates within us is powerful enough to fill our mental hospitals with people who see and hear things that aren't there. Among the sights and sounds that originate within us are our images of ourselves and our realities—our archetypes. Such images are powerful icons, nearly immutable. These are the images of our dreams, our poetry, our theaters, our psychoses.

If physical reality, the outside world, changes abruptly, it may not be within our power to so abruptly change such deep-rooted images of ourselves and our worlds. When that happens, reality itself becomes implausible. Then our only way out is through a phantom, a bit of virtual reality that reconciles our world and the real world.

Are the dead, then, living within my neurons—inside of my own pictures of me?

Images of ourselves—some, apparently, older than we are—are obviously deeply etched into the stones of our minds. Powerful things that resist change, particularly sudden change. But even these archetypal portraits of ourselves aren't without seams or cracks. And inside those seams and between those cracks, small forces working over years can introduce change. Time, in an intimate and powerful relationship, could reshape even our images of ourselves. The changes would be little ones at first, a tiny fissure unmortared here or there, room to include in our self-portraits parts of other men or women, a first vision of ourselves as something more. Later, larger pieces of us might be lifted and replaced by whole chunks of another. Husband and wife begin to speak alike, know what the other is thinking, anticipate what the other will say, even begin to look alike. Until one day, what remains is truly and thoroughly a mosaic, a chimera—part man, part woman; part someone, part someone else.

And then, if that man or woman is amputated from us, clipped as quickly and as cleanly as a gangrenous leg, our minds are suddenly forced into a new reality—a reality without the other, a reality in which an essential piece of us is missing. At that point, our declarative minds would be at odds with our own pictures of ourselves. To rectify that, to reconcile the frames flickering inside with the darkness flaming outside, we conjure a phantom, a phantom to change our worlds. We force a bit of what is inside out there into the real world, to create someone or something that will help us slow the universe for a moment while we repaint our pictures of ourselves with a very small brush on a very large canvas.

Today, sitting on the redwood deck behind my house, the air smells of cinnamon and rainwater. For reasons I can't recall, those smells remind me of the Brandenburg Concertos, coffee on Sunday mornings, and the intricate paths of swallows.

Somewhere inside of me, there is a woman. But where she lives and who it was that led her into that pastry shop last Thursday, I've no way of knowing. For one part of me, that ignorance is a gnawing blindness. For another part of me, it is enough to simply know for certain that I will see her again.

*GERALD N. CALLAHAN is an immunologist in the Department of Pathology at Colorado State University in Fort Collins. His first book,* River Odyssey, *is a collection of personal essays and poetry (University Press of Colorado, 1998).*

# Proud Flesh

MARA GORMAN

In February my mother dreamt of moving to Santa Fe, where the nights are cold and each day is brilliant, dry, and full of possibility. Waiting, she sat on her bed under a cabbage rose comforter, her bald head encased in a lilac turban. At times, she listlessly read romance novels and self-help books. But mostly she lay flat, eyes turned to the ceiling, imagining her southwestern apartment that would have, as she told me, "two floors, a loft bedroom, big windows, lots of light, sliding-glass doors, and a garage." Her eyes were huge and shadowed when she said this, and then she turned to me and asked, "That doesn't seem a lot to ask does it? I mean, to have a garage?"

The chemotherapy made her nauseated, the antinausea drugs made her woozy, and the wooziness made her cross. Weary of illness after the tedium of two months, she thought it was time to be better—a big joke between us: my mother was an *im*patient. So I smiled and offered water, Mint Milanos, lemon-scented tea, crusty Italian bread, carnations, Godiva chocolate, all of which she rejected. Then I went into her small bathroom and picked wispy remains of her hair out of the bathtub drain. I looked in the mirror at my bare face and tried to imagine the shape of my mother's skull; she would not yet let me see her bareheaded, and I was one of the few people she allowed to view her in only a turban or hat with no wig. I stared at the hollows under my eyes, and I did not see my mother, who is like the antique dolls she loves: luminous, but fragile, chipped, and slightly sullied. My face revealed nothing to me, no fissures, just smooth skin, pursed lips, carefully sculpted eyebrows.

I was in New Jersey that winter of 1996 to take care of my mother. She lived alone in a small apartment and preferred that I drive four hours from my house in Pennsylvania to clean her tub rather than accept help from her friendly and cheerful downstairs neighbor, or her sister-in-law, who lived five minutes away. Of course my mother did not ask me to come, she would never impose herself on me that way. She simply accepted the offers of assistance we both knew I would make, part of our silent game whose rules dictated that she not ask me to be responsible, knowing all the while I would be. Or perhaps she thought that if I paid the debt of eighteen years of sacrifice, I might escape a fate similar to hers: to be ill, to be poor, most of all to be single at a moment of crisis. And I wanted to be patient, nurturing, loving, and amusing—Florence Nightingale and Jo March rolled into one. Convinced that I just needed to find the right degree of cheeriness, the right amount of love, the proper incantation to talk over my mother's bed and erase all suffering, I drove to New Jersey as often as three long weekends a month and did not complain.

The atmosphere in my mother's apartment was both oppressive and somehow comforting, reminding me of our house when I was a child. The almost tangible thickness of the quiet, the way the sound of words could be absorbed was very familiar. Childhood images are for me a pastiche of muffled sound. Silent summer evenings, heat lightning shimmering, walls threatening to close in on my graceless, chunky seven-year-old body clenched in my bed awaiting thunder that never came. My voice, always carrying more than I meant it to, always saying more than I meant it to, rising to meet my mother's soft tenor, as she told me to please try not to speak so loudly. A shriek in the yard as I pull my sister Sheila's hair, quickly squelched by my mother's soft scolding. These images blend easily into the night of my mother's mastectomy, when in the dark silence of her apartment I dreamt that all of my teeth had fallen out one by one, each tooth coming loose with excruciating slowness, my tongue pushing it around before it fell, plop, to the floor next to my chair. When I woke in the morning in a mass of drool, my head throbbed and I jumped up and turned on the radio just to know sound.

I viewed my mother's illness as holistic, as it seemed to me a result of years of simultaneous struggle and quiescence. Life had long been something to overcome for her, something to get through day by day, small dabs of joy arriving irregularly to be accepted, even welcomed momentarily, before a return to the routine of uncomplaining difficulty. I was eight, Sheila six, when my father left; we saw him and my stepmother often, but lived always with our mother. Thinking about our family, both before and after the divorce, I imagine eighteenth-century Dutch interiors: still and profoundly silent rooms, muted colors. While the rooms were full of light, the atmosphere was clogged with words that no one heard, suspended in the heavy air. I imagined that in the struggle to not feel, to not hear, the many muffled words had turned into tumors, growing in my mother's chest. Appointed by myself and my mother, I struggled to free these words and bind old and new wounds, thinking it would save my mother's life.

I picked up the phone one October evening to hear my mother's low, warm voice: "I'm going in for a biopsy tomorrow. I've found a lump." She had known for two weeks but hadn't wanted to worry me; if the surgery had not required anesthesia, I doubt she would have told me at all. Of course I did worry, for about an hour and a half, but my mother was so calm, so convinced that this was nothing, an irritation, a mysterious build-up, a woman thing, that the worry took no more shape than a tug at the back of my brain. After all, I was only twenty-five. But that moment when the fear first entered my mind was the beginning of a long, slow, complicated dance, during the course of which my mother and I switched roles.

The moment when we receive bad news should be dramatic, but is really often mundane; we are aware of the phone cradled against our shoulder, of the cuticles on our hands, of a dull ache at our temple. The universe shifts and we stand there stroking the kitchen counter.

The phone rang again two days later. "The doctor just called and I have cancer. I have cancer in my right breast." I stared out the window thinking everything will be different now; nothing can be the same after this moment. Then I cried. But I did not truly understand what was happening; I went through the motions of distress, through the rituals of sorrow and disbelief. Sheila came over to my apartment and we attempted to smile, painted each other's fingernails and cried over a video of *Little Women;* Winona Ryder was supposed to be Jo but was far too delicate and pretty, and besides Jo never kissed Laurie. We noted these changes without enthusiasm. Mostly, we did not talk.

The next day I went to the library and took out a stack of books. Here is what I learned: My mother's cancer was a type usually labeled a precancerous condition, an isolated tumor easily removed, but since it had already started to spread, it was no longer precancerous, but full-fledged invasive cancer. The doctor did not know to what extent, and where, if at all, beyond her breast tissue it had radiated. No neat and tidy tumor for her. I read that a single cancer cell can take up to a year to split into two cells; a two-millimeter tumor is made up of millions of cells. But sometimes cells split more quickly; no one knows why. The chances of surviving breast cancer are 40 percent, except when they are 10 percent or 60 percent. No one knows exactly how chemotherapy works, but they do know it kills every rapidly splitting cell in the body, which is why it causes the patient's hair to fall out—it kills the hair cells and white blood cells as well as cancer cells. Cancer can be controlled by diet and positive thinking except when it can't be and it spreads everywhere. Breast cancer is always breast cancer no matter what organ it chooses to attach itself to in the future. My chances of getting breast cancer at some point were now double what they used to be, a fact repeated to me by every health professional I encountered in the next month.

I read the books for as long as I could bear, and then I brought them to New Jersey for my mother to help her make rational decisions about what savagery she would have them do to her body.

Two weeks after my mother found out she was ill (with cancer, of course, one often has to be told; my mother certainly did not feel sick), and ten days prior to the mastectomy, I sat in the azure-carpeted office of her plastic surgeon staring at his degrees: one from Emory, two from Rutgers, residency and chief residency at the University of Cincinnati Hospital. My mother was in a room somewhere having photographs taken of her breasts while I studied the gleaming mahogany surface of his desk and read a pamphlet that described how he could remove tissue from her stomach to replace her breast, "just like getting a tummy tuck, too!" he had said with a laugh. This was my first trip to visit my mother since she was told she had cancer. I had already promised to spend the week of the surgery with her and to rearrange my life in whatever ways necessary to ensure she got the care she needed. She was not gracious about accepting these offers and tended toward vague assent. But the way she had taken my hand in the waiting room fifteen minutes before, the coolness and weight of her fingers pressing into my palm, was enough to convince me that we both needed me to be there.

Upon hearing the news that one of her breasts had to be removed—and quickly—my mother had fixated on the idea that the mastectomy would at least provide her with the opportunity to have larger breasts. She decided almost immediately that she would have reconstructive surgery, a procedure that she thought would render her more "feminine," a term I was still struggling to understand. To me she was, and always had been, beautiful. But her breasts were a source of shame to her; she considered them so small as to be deformed. A teenager in the fifties, her image of what breasts should be was influenced by the pointy missiles of Marilyn Monroe and Jayne Mansfield.

I never saw my mother's breasts except perhaps as a small child when we bathed together, nor did I think about the fact that although she is three inches taller than me, her bra size was a whole cup smaller, or that her bras were all padded, or that she never wore tight sweaters. She was thin and tall, which I always envied. I also thought she cared little about her appearance: she never read my fashion magazines, never had her hair colored, wore little makeup, didn't care about clothes. In fact, when my parents were still married, she wore mechanics' coveralls every day to paint in, her only perfume Jean Naté and turpentine. But her hair was always soft, her hands always graceful and strong.

To me, her breasts simply existed; I attached no specific importance to them. But to my mother they represented a lack of sexiness. She wanted no one to touch or see them, and as I was to discover, had until this point refused to have a mammogram. During the weeks that preceded the surgery, my mother, while she talked quite a bit about the pros and cons of breast implants, did not really discuss the shame and fear that had caused her to neglect her breasts until the lump was big, a calcified ridge hard to the touch. She invited me to stroke it, to feel how unyielding it was, and I did, not asking why she had avoided having mammograms.

When my mother and the doctor, who has an improbably open and young face contradicted only by slivers of gray in his dark brown hair, came back into the room, he immediately started to lay out her options. Although cancer was only detected in her right breast, insurance would pay for both breasts to be removed as a preventative measure. To do so would eliminate the risk of the cancer appearing in her left breast, the chances of which were anywhere from 10 to 40 percent. It would also mean that insurance would pay to have matching implants inserted so that her breasts would be of equal size. My mother could opt to have temporary implants put in at the time of her mastectomy, which would mean that she would not have to go under anesthesia twice, or she could plan on doing the reconstruction later. If she chose the former, she would return to the doctor's office once a week for a month after the surgery to have saline pumped into the implants, stretching the tissue (despite the ominous sound of the word "stretching," it did not occur to me or to my mother that this process would be painful) and preparing her breasts for the permanent implants, which could be put in six months later.

I sat, taking careful notes, glancing at my mother, who was trying to look thoughtful, but whose distress was evident in the fixed look she now gave the doctor. "Don't make a decision now," the doctor said, "think about it for a while."

My mother had ten days to decide.

In the car a half hour later, she talked through her decision. "You know," she said, "this doesn't have to be all bad. I could end up with a nice pair of breasts."

I started to giggle. "A real pair of boobs," I agreed, "huge boobs." She started to laugh too. "A matched set," I said, "real bazongas. You could look like Barbie."

"But seriously, Mara," she said when we were through laughing. "I really could end up with nice breasts."

I didn't tell her that I thought her breasts were nice then, but simply nodded my head.

The evening before her surgery, my mother donned pink satin and danced until her legs hurt. Driving to the restaurant where her dance group was having a dinner party, we did not discuss what would happen the next day. I spent the weekend trying to be funny and cheerful, and my reserves of perkiness were running low. I was afraid and could not help staring at my mother, as if to memorize her body.

Ballroom dancing is, besides reading, my mother's only escape, her passion. She took it up when Sheila and I left home and is now very good at it. Her seventy-year-old teacher, a grizzled man who cannot see well and who spills food on his tie when he eats, thinks he is in love with her. "Your mother is so beautiful, see how beautifully she dances?" he would breathe in my ear every time he was not out on the dance floor with her. Then he would ask for the next dance. If it were a cha-cha, she danced with absolute abandon, hips swaying, smiling, a more sensual being than I have ever seen her in my life, absolutely at home with her body. Watching her glowing face, the smoothness of her feet, I wanted her to stay whole. I had never seen her look so unmarred.

The day of the surgery, the Monday before Thanksgiving, dawned flawless and bright for November. My mother rose as usual at six o'clock and showered, putting on eye shadow and mascara and pearly lipstick. She drove us to the lab where she would have blood work done, and we made lame jokes about not having coffee. Her hair was caught up on top of her head in a pink scrunchie and she looked young.

I had placed a picture of me, two years old, wearing soft, footed, pink pajamas, holding my infant sister, who is nothing but a head and a bundle of blankets, into my mother's overnight bag, along with a plastic angel playing the violin. Ritual offerings to the altar of her suffering, this was the first of my attempts to cure my mother with objects. I wanted to believe that if I surrounded her with beauty and goodwill, she would get well. I was not alone in this desire: flowers, balloons, and stuffed animals arrived daily for weeks after her surgery, and I arranged them around her bedroom in an attempt to make it seem cheerful.

Sitting tensely in the grim, windowless waiting room in the bowels of the hospital, I held the overnight bag, my mother's purse, and my own bookbag, while my mother disappeared behind a forbidding door that wore a sign saying "Authorized Personnel Only." The waiting room was hardly larger than a closet. Crammed with vanilla-colored plastic chairs that were attached to bars on the floor, it smelled of fear and antiseptic. A girl of about sixteen, lipstick and fingernails black, clutched a teddy bear and silver Mylar balloons that said "Get Well Soon" and "We Luv You." She slouched and scowled, Mariah Carey squeaking out of her Walkman. Her companions were a young couple, both with stringy hair and soft bellies that spilled over the waistbands of their jeans. They did not speak but sat, shoulders pressed rigidly against the back of the chairs, occasionally shifting their weight. None of us looked at each other; I stared at the poster of pink tulips that hung over their heads, they at the floor. When a child's screams became audible in the recovery room next door, they leapt to their feet at the same time in a smooth, swift motion, as a nurse emerged to usher them to their son's bedside. The girl shuffled after them, singing along to the music. I felt the cold plastic through the thickness of my coat. There was no way to be comfortable in these chairs.

A nurse came out to fetch me once my mother was robed and lying on a gurney with an IV attached to her hand. The nurse invited me to follow her as she wheeled the gurney into the holding area, an unexpected invitation, and when my mother nodded her assent an orderly quickly forced a hospital gown over my coat. Bulky and burdened with all of the bags, I staggered into a large, dim room, a meat locker filled with stretchers and silent green figures. A room from a dream, where I sat beside my mother who was thin, calm, and pale. The anesthesiologist, a small, bearded man wearing scrubs, hovered over her as I sat in a chair stroking her long, cool fingers. Next to us, a young woman in sweatpants and basketball shoes dabbed her eyes as she clutched at the hand of what looked like her grandmother, a gentle-faced man of about sixty, wearing a red-and-black-checked flannel shirt sat across from us holding onto his son, prone like my mother. All energy in the room was focused on those lying inert, as though the simple act of leaning over their stretchers could cure them.

My mother's scrunchie lay against the pillow. I tried to memorize her exactly as she was and not imagine where she would be going, which was difficult given that people kept coming in and out of the swinging doors behind us through which I could see the silver surgical lamps that would hover over my mother during the operation. A red-haired nurse came over and introduced herself, promising to call me when the surgeon finished cutting the breast tissue off to tell me how things were going.

My mother had decided to have both breasts removed, leaving only the nipple of her left breast intact. As a result she would be undergoing two procedures that day: First the surgeon would remove the tissue from both her breasts, as well as a number of lymph nodes from the right side. Then the plastic surgeon would insert the temporary implants, limp plastic bags, that he would gradually fill with saline over the next several months—just like blowing up a balloon, my mother and I joked uneasily. Mastectomies have become so common that this surgery is no longer considered "major"; some women are sent home the same night. But by adding the plastic surgery, the procedure would take more than twice as long, close to nine hours in my mother's case, although we did not yet know it would take so long.

My mother urged me not to sit in the waiting room. "It's such a lovely day," she said. "One of us should enjoy it." When they finally made me leave, they were wheeling her into the operating room. She gave me a last kiss, tightly puckered as her kisses always are, and disappeared. Rather than leave and miss the opportunity to speak with the doctor, I decided to stay in the waiting room until the surgeon emerged; then I would buy some lunch and wait for the plastic surgeon to finish his work.

I paged through magazines, women in evening gowns gazing back impassively, Sharon Stone in a thousand-dollar cashmere sweater—my mother looked fine, felt fine, and yet her body was host to millions of multiplying cells that could kill her. Facials protect winter-ravaged skin; the key is moisture, moisture. She was being sliced open, made to bleed, scraped clean, and sewed up. The glossy paper slipped through my fingers, the pictures whispered seductively. When

I saw her next, she would be missing chunks of her body. Gleaming pictures, gold for the holidays, sequined gowns, velvet, silk, the textures of desire, the curve of perfect breasts covered in red satin.

Although I had not had breakfast, I ate nothing and ultimately could not read, could only stare at the cinderblock walls. I was called up to the glassed-in nurses' station around noon to speak on the phone with the red-haired nurse. I could hear the beeps and clicks of machines monitoring my mother's vital signs. "Things are going well," she said, "but it's taking a little longer than we expected. Dr. L. should be out around one o'clock." I returned to my seat in the corner, waiting.

It occurred to me then, for the first time, that if something awful did happen, if my mother died or slipped into a coma, I would be in this barren hospital all alone. I was the husband, the mother and father, the sister and the faithful daughter, and I was expected to comfort myself, for the nurses made it clear that they cared only that no one make a scene. At the slightest hint of distress or panic, they spoke in clipped tones, telling a young woman's husband that there was no way to know when she would be out of surgery; no one can predict these things, can they. Calm is of course impossible in a place where everyone is waiting with balloons and flowers and guarded expressions of studied nonchalance on their faces, and when one o'clock, then two o'clock passed with no sign of the surgeon, my agitation grew unbearable. Tearful calls to my boyfriend and father and stepmother did little to satisfy me.

At three o'clock, a short woman of about sixty, with patently fake red hair marched into the waiting room and announced that she was a volunteer and would do her best to get us information, but not to expect too much and to realize that we weren't doing our loved ones any favors by worrying. I hated her on sight. By now, I was in a state of blind panic, convinced that something had gone horribly wrong. I had no choice but to ask the volunteer, who stared while I explained my concern and before I could even finish replied, "You can never tell with these doctors. Things just take longer than they say. Why don't you go and sit down." By now, someone had turned on the television to a talk show where a woman screamed at her husband that she slept with his brother because the sex was so much better. The audience oohed and hooted. No one really watched, and I finally turned it off when the volunteer told me that she had checked, and my mother was still in surgery. When she told me not to worry, I thought of a million possible replies, but only smiled wanly.

At six o'clock, the plastic surgeon came out. Oh yes, Dr. L. had finished up four hours before (to this day, I do not know why he did not come out to speak with me). I was limp and cold, trembling with hunger and fear. My mother was fine, but she had lost a great deal of blood, and the surgery had taken so long that they were having difficulty waking her from the anesthesia. I was not allowed to see her. "Why don't you come back in the morning?" he said, and patted me on the shoulder before walking away.

When I entered the hospital room the next day, my mother was asleep. She woke as I stood looking at her, and moved her lips as though she had no teeth and was trying to gum a mouthful of food. Her chest was swathed in bandages, tubes running out of the sides. The picture and angel, which I had placed in her room the previous evening before leaving to go home, were lost in the clutter of paper cups on the movable table next to her bed. She could not smile, could only sigh and moan and sleep.

Happily, she had the room to herself, and I spent the day curled up in a chair at the end of her bed, staring at the still yellow and orange leaves out the window. The hospital was warm, and occasionally I dozed as my mother drifted in and out of consciousness to poke unhappily at her food or request more ice water. Nurses entered the room at infrequent intervals to check her blood pressure or empty her drains. They would always leave without adjusting the pillows, abandoning my mother to gasps and trembling as the pain of her wound set in. I learned quickly to fix the pillows so she was comfortable and to keep her pitcher full of water. Her legs were attached to a machine that vibrated and hummed, preventing blood clots from forming. Even though she had only woken from her surgery hours before, and was offered a bed pan, she refused to use it. But she could not reach down and unhook her legs from the machine I mentally called "the Vibrator," and she had very little bladder control, so I learned to jump up as soon as she woke to help untangle her IV and detach her so that she could go to the bathroom. I cut up her food and stroked her hair.

When I returned to her apartment that evening, I feverishly cleaned her room, sweeping piles of paper off the table, arranging flowers and get well cards in a frenzy until I collapsed on the twin bed, pinned by my mother's pain, unable to move.

Blood, the most extraordinary crimson, filled four plastic drains, attached to my mother by tubing that disappeared under the bandages covering her upper torso. We marveled together at its deep rich hue, "Have you ever seen anything so beautiful, Mara?" my mother asked, "It's remarkable." Four times a day, for five days after the surgery, I carefully squeezed the blood into small cups, like the ones used to dispense cough syrup to children, and carried it into the bathroom, where it flushed down the toilet in ruby swirls.

So passed the next week. I exactly followed the regimen set by the doctors, made sure my mother took her pills at the proper times, emptied her drains, tidied up her bed, kept her cups full of fluid and made her drink them, prepared meals and coaxed my mother to eat, and did dishes. When my sister arrived, we shared the duties, but I supervised. No detail escaped my watchful eye.

My mother quickly grew resentful. Why don't I feel better? she would ask, her face drawn and pale. Even the get well cards and chocolates did not make her smile. I tried to be patient, sometimes succeeding. You lost a lot of blood, I told her, you are still losing blood. You have been through hours of surgery. You have a foreign substance in your body. I want to be better, she replied stubbornly, and all I could do was give her more painkillers.

Early in the morning on the Saturday after the surgery, I again visited the plastic surgeon's office with my mother. She asked me to come into the examining room to wait with her for the doctor. I sat on a stool and held her purse in my lap. When the doctor came in, I expected my mother to tell me to leave, but she said nothing. He looked at my careful lists—four times a day I had recorded how much fluid came out of each drain—and smiled at me. "You've done a good job," he said. He told my mother that he thought two of the drains could come off and asked her if she wanted me to leave while he unwrapped the bandages. To my surprise, she said no.

He and his nurse started to unwind the Ace bandages and gauze that covered my mother's chest, round and round, like something from a bad horror film, the nurse's curved magenta fingernails pulling gently at brown-spotted gauze; my mother's face was calm, expectant. Underneath the gauze, her skin was surprisingly rosy. One side was flat, nippleless, a curved black scar circling underneath her arm. Her skin looked soft. The nipple remained on the other side, pink against her chest. Like a child's breast, there were no curves, just the ghost of womanhood in that one remaining nipple.

The nurse kept looking at me, but I remained unperturbed until the doctor cut the tubing to which the drains were attached. My mother made short gasping noises that were worse in their restraint than shrieks of pain would have been. I winced and the nurse, in a voice full of concern, asked if I was all right. But I was fine. My mother's scars did not scare me, but the sound of her pain was shocking.

In the car, later, after I argued with her to let me carry her purse and drive us home, my mother leaned her head against the seat and closed her eyes. "That wasn't so bad looking, was it?" I agreed that it wasn't. "I wanted you to see, Mara, in case you ever have to go through this too. I wanted you to not be afraid. It's not bad, really, there's a kind of innocence about it." She sat up and looked at me.

"Don't you think? A kind of innocence?"

They found five tumors in my mother's breast, two of them more than a centimeter across, which is enormous in cancer terms. The size of the tumors was enough to necessitate five months of chemotherapy although no cancer was found in her lymph nodes. The doctors simply could not be certain that the cancer had not spread. The reconstructive surgery proved to be enormously painful. Each of the four times my mother had to have the "stretching" injections, she could hardly move for days. When at the end of her chemotherapy, she was to have the surgery to place the permanent implants, it was discovered that one of the temporary implants had torn, leaking saline and causing an infection. Once again, they opened her up, scraped her clean, put in new implants, and sewed her shut. My mother's breasts are now much larger than they used to be.

Through this all, I drove to New Jersey, went to the grocery store, and cleaned the apartment. I listened to my mother's grief and loss as she grew weaker from the chemotherapy, as her temporary implants became stiff and hard (when I had joked about Barbie breasts, I had not realized how prescient this was), as she slowly lost her hair and her eyelashes. Her denuded body, she told me, looked like a child's. "When I look in the mirror, I feel vulnerable," she said.

By February, when I emerged from the shower, soft scrolls of my mother's hair covered the front of my chest, forming gentle brown curlicues across my unscarred breasts. Picking them off carefully, I placed them on the top of the brimming garbage can. I vacuumed before I left for home, and emptied the trash, did the dishes, organized the get well cards and pill bottles, tried to remove the detritus of illness.

Pieces of my mother's hair were everywhere in the stuffy apartment and I had nothing to offer but my housewifely skills and my ear;

I tried to listen to my mother and to be patient. Often I failed, growing exasperated with her passivity, "Won't it be great when I'm better, Mara? Won't it be wonderful when I feel good?" I wanted her to live in the present, to accept what had happened and therefore recover from it. This was especially unreasonable because while my mother has survived every wound that life has inflicted on her, no one I know carries around more scar tissue than she.

And so I looked in the mirror and denied that I saw my mother's face reflected back in any way. I like to think I would do things differently than my mother, that instead of devastating my life, cancer would be a project, a challenge. My list would include the following: buy lingerie, learn a foreign language using tapes, invite my many friends over to have tea and chat, figure out an anticancer diet and follow it to the letter, read every piece of literature and not permit my doctors to make decisions for me, not replace my breasts with implants, display my bald head proudly.

The reality, I know, is that I would probably sit there, as my mother did, in the middle of all my struggles for perfection in the face of disaster, and wait for the moment of happiness. I would grow tired of the discomfort and cry for the slow departure of my hair, the mutilation of my body, the loss of my femininity. Imagining my mother's pale face and smooth head, I know that I, too, would wait.

***Mara Gorman** recently graduated from the Pennsylvania State University with an MFA in nonfiction writing. She is an assistant editor at the International Reading Association.*

# Power

SUSAN L. FELDMAN

In 1986, I began to make monthly trips from my home in San Francisco to consult with a doctor in Baltimore offering a new, radical, and promising treatment for the autoimmune disease that was slowly but steadily swallowing up my life. All told, I made this trip eighteen times. Since I had to have my blood drawn before the treatment—and then once again the morning following—this necessitated my staying at least overnight. Having been ill for a number of years, I had already learned that the only way for me to play the hand I was dealt was to maintain the illusion of always seeming to seize the upper hand. Thus I became determined to find a way to do the whole trip to Baltimore each month in forty-eight hours and lose as little more of my life to illness as possible. As there were no nonstop flights to Baltimore, I took this on as something of a geographic challenge.

I began with the coast-to-coast run and then the quick hop from Kennedy to Baltimore, but my bag got lost and the change of planes required a sprint that was way beyond my abilities. I then tried the three-hour stopover in St. Louis, but found the airport smoke-filled and without a quiet place to sit and productively pass that time. So it was with not a little triumph that in the fifth month I was able to perfect my run and make it home in just under two days: the red-eye from San Francisco to Chicago's O'Hare arriving at 5:11 A.M. local time with a change to a flight leaving at 6:30 A.M. and arriving 9:09 A.M. at

Baltimore airport. Dashing into a cab, I could check into my hotel and still be in the doctor's waiting room at ten o'clock sharp. And by the third time I did this, I found a quiet place in O'Hare where I could sit with a cup of tea and watch the planes taking off into the early morning sky while I awaited my own flight.

But the eight hours of travel time remained empty space that I could not fill. The disease had struck first in my eyes some seven years before, and I had been left with a constant inflammation that made reading difficult, usually impossible. As I write these words now more than a decade later—and having recovered most of my vision—it astonishes me still to watch with what ease they roll onto the screen clearly in my sight. On a November morning in 1979, the ophthalmologist had shut off the slit lamp, leaned back in his chair and looked away from me. "I guess you won't be able to read, will you?" he said. "What do I do?" I asked, not really capable of hearing an answer but trying instead to stave off the tidal wave of terror that had arisen at this confirmation of a truth I had long suspected. He shrugged. "Learn Braille I guess," and with that he stood to indicate the visit was over and I was to leave. I walked out then into the bright sunlight of Fifth Avenue knowing I hadn't the vaguest idea of where to go: as a teacher and writer, my whole life had been made up of moving from text to text.

The first months of visual impairment were an abyss of disbelief, rage, impotence, and despair. Though I could seemingly engage in normal conversation, the roar in my head at the blur in my vision made everything almost surreal and absurd. I spent that time doing literally nothing but shaking my fist at the gods who had taken away from me the one thing that had always mattered most: words on the printed page.

I found my way out of that overpowering sense of blackness through the curious course of signing up at a gym and starting to lift weights. Having spent months with nothing to do and no place to go each day, I had become sluggish and heavy. Finally, I decided to go to my local YMCA to swim but instead signed up for a Nautilus class after I heard two women talk about it in the locker room. In the gym, I found myself intrigued by the high-tech equipment and the orderly progression of the circuit. Never having been particularly "athletic" before, I undertook this weight-lifting program with the completely conscious intent of mastering my body—by force if need be—into doing what *I* now wanted it to do. It became the necessary and sanity-saving anodyne to the progression of the disease in my eyes.

As my body grew more powerful and I became aware of how my efforts had produced muscles that could lift nearly my own weight with ease, I began to ask myself how I might direct this power into approaching the central problem of my life: how to read and write when I could no longer read nor write. Soon I returned from my daily visits to the gym and sat on my couch with the phone planted in front of me and began to make calls. After weeks of frustration, I worked my way through the various city, county, state, federal, and private agencies and was finally able to line up the institutions and resources available to me. Through them, I acquired taped books and volunteer readers.

At the suggestion of one social worker, I selected a book I knew well and listened to it over and over and over with my eyes shut till I no longer heard the reader but merely the words—*Pride and Prejudice*—my gateway to the printed page once again. I became accustomed to using tapes and cassette players, to carrying a small microcassette player with me at all times in lieu of pad and pen. I advertised at the local university and hired a graduate student as an amanuensis while I studied the history of Milton, Henry James, Borges, Sartre, and Dostoyevsky, all of whom had to confront for one reason or another the obstacles of "writing" by dictation. I became an excellent "tape reader," able to listen at three times the normal speed. What sounded like Donald Duck to others came to be the equivalent of the speed-reading I'd done all my adult life. I joined organizations for the visually impaired and disabled and began to learn Braille toward the day when the worst-case scenario could occur and I might need it. Fully-sighted people commented to me that learning Braille must be like learning a new language. But it was nothing like learning a new language; it was like developing a new sense, and I found it daunting and elegant and readily accepted I'd never be much good at it.

All of this, however, was much behind me by the time I began regularly boarding the midnight flight from San Francisco to Chicago each month on my two-day trip to Baltimore and back. But in spite of having the best equipment and years of experience, I could not "read" taped books on the flights. The tapes, produced primarily by Recording for the Blind, would become distorted beyond recognition if turned up loud enough to be heard above the plane's engines. Month after month, I tried better earphones, better players, even boosting the tapes by re-recording them, but nothing worked. All I got was the steady drone of the engine beneath harsh, ratchety voices—the phrasing of John Updike, Barbara Pym, and Mrs. Gaskell lost in the electronics. I took instead to knitting during the nights when I flew and furiously tapping my foot at the time wasted, time lost.

In the fourteenth month, I sat on the runway at Baltimore-Washington International waiting for my return flight to take off. It was summer and I'd found it hotter and stickier in Baltimore than it seemed dignified for humans to endure. Because of some delay, the plane's air-conditioning hadn't been switched on yet. I was past suspecting and now certain that this promising new treatment was having little effect to the good, and whatever charms the Walters Art Gallery and National Aquarium had offered in my monthly visits to Baltimore had long since run to the thin. At that moment, I wanted to be going forward, to be moving aloft, anything but considering being stymied once again. I wanted all the voices in my head crowding in to remind me that not only weren't there any easy answers—but there weren't any answers at all—to be silenced for a time till I could be home, in the fog and the cool and the familiar. But the pilot was announcing some kind of new delay.

The flight attendant came around with complimentary earphones in an attempt to mollify the restless passengers as the time passed and the palpable heat and impatience increased. My seat companion had immediately informed me that he was an avid golfer and was now asking me about my favorite in an upcoming tournament that I knew nothing about. Nodding and smiling, I put the earphones on and flipped on the preprogrammed music. Unable to read the inflight magazine, which would have told me the program, I simply kept turning the dial until I came to classical music. What I heard was Mozart's Piano Concerto no. 21, which I knew had been used as the theme in the Swedish film *Elvira Madigan* and had been called that ever since; I resigned myself to hours of "Great Movie Hits from the Classics" and began to further curse my fate.

At about the time the plane was leveling off along with the temperature, the flight attendant came round to ask about drinks and I heard the opening of a piece of music that I did not know though could easily identify as Beethoven. At first, I couldn't even figure out what kind of work it was—surely orchestral with a piano but it was not any of the concertos that I knew. I began to listen intently and at times felt that I was hearing what seemed to be variations or echoes or something of what I knew to be passages of the Ninth Symphony.

Then there was a great build, first with the repetition of the violins, and then movement up the keyboard of the solo piano. This was followed by a passage of tender, coaxing beauty and everything around me began to melt away as I heard and saw and knew nothing but the music coming from that piano and those violins.

When the chorus began to sing, I was practically lifted up out of my seat with surprise; I hadn't expected voices—why should I? I thought I was listening to a rarely performed concerto for orchestra and piano. But suddenly women's voices filled my ears, singing in German a song, a lyric, a poem I could not know. In counterpoint to the piano, their voices suddenly encompassed the world for me.

I closed my eyes and began to listen very very carefully, the early years with my Yiddish-speaking grandparents giving me some small ear for German. But all I could hear then was the piano as it wound its way through the passages and the chorus sometimes against and then over it. Like the fourth movement of the Ninth, this work also teased the listener along with moments of great stillness that would then break out into measures that were building blocks to a higher and higher plane, keeping you always aware that you were being led to some extraordinary climax that the composer alone could envision.

And then I heard a phrase over and over, words I knew and could almost understand: "Lieb und Kraft! Und Kraft! Und Kraft! Lieb." I of course knew, my face having been cupped a thousand times by elderly hands as I was called "love" for the tiniest thing I did or even for nothing at all. But *Kraft,* surely I must know this word, I kept thinking. If I just listen, I'll be able to get it. But I couldn't.

The music continued to climb and then seemed to seize on that word and take flight upon it. "Und Kraft! Und Kraft! Und Kraft!" the chorus sang. And . . . and . . . and what? But then it was impossible to find the word in my memory as I was completely caught in the music as it reached for its grand end. Four great arpeggios. And it was there, there in the final chords of the work that I found the meaning at last: *Und Kraft!* As the music surged up the keyboard, I felt again that moment of the last lift of the weights above my head, 120 pounds held, my arms taut, and strong, and free: And Power! And Power! And Power!

After the last chord, I turned to the window though there was nothing but clouds to see outside; my face was streaming with tears and I didn't want the golfer to ask why. What would I have said in reply: live my life these last eight years and then listen to this piece by Beethoven? For the rest of the flight, I stayed attentively tuned to the classical channel knowing the Beethoven would come around again and again. By the time we landed in Chicago, I'd heard it three times and had torn the pages out of the in-flight magazine that gave the program listing.

When my grad student came the next day, I asked her to read me the classical pieces listed in the program till she came to Fantasia for Piano, Chorus, and Orchestra in C minor, op. 80. "Does it give the year?" I asked. She shook her head. I led her down to my living room and had her check through the index of Beethoven's collected work. During my well and young years, I'd purchased the *Deutsche Grammophon Beethoven Bicentennial Collection* and my assistant now got out Volume VI, Choral Music. "There's a booklet inside," I said. "Find the section on the Fantasia and see if there's a date given for the piece." She found the pertinent paragraphs and read them aloud to me. As she did this, I took the record out of its sleeve and put it on. While the opening moments held no surprise, even expected recognition could not diminish their power.

"There's a translation of the poem here, do you want to hear it?" my assistant asked.

"Yes," I said. "Who wrote the poem—it's not Schiller, is it?"

"No, it says Christoph Kuffner, but there's a question mark after his name."

"Must be attributed to him but I guess there's some dispute. We'll look it up in Thayer's biography later. Read me the poem . . . slowly." Just as the chorus began to sing in the German she came to the last verse:

> The greatness which permeates the heart
> Blooms again with fresh beauty.
> When the spirit exalts,
> A spirit chorus reverberates for ever.
> Then take with joy, o noble spirits,
> The gifts of high art.
> When love and power unite,
> Almighty grace endows mankind.

"Listen," I said, "listen carefully to what the composer does with the word *Kraft*. It means power. Listen to how he uses the word and the music to create his meaning." Together we listened through to the final crescendo and the piano's triumphant chords, extraordinary—exultant.

I sat silently for several seconds until the skipping of the needle at the record's end made me get up. "Beethoven wrote that in 1808," I said.

"What's significant about 1808?" my assistant asked.

"Nothing. There's nothing significant about 1808 except that Beethoven wrote this piece in 1808 . . . and that by 1799 he must have already suspected he was going deaf and by 1808 it was a certainty he could no longer ignore," I said, putting the album away.

Nearly a decade later, I am sitting in the same living room with my husband listening to our local classical music station. My husband has been trained as both a pianist and classical guitarist, and I know that he listens to music as I read fiction, attentive to layers of meaning and device. I look at my watch. "They're going to play the Beethoven Fantasia soon."

"How do you know?" he asks.

"They have a feature now where listeners get to suggest pieces. I e-mailed in the Fantasia. I'm sure they'll play it."

He smiles. After only two years of marriage he is still charmed by my certainties, whether appropriate or not. "Why?"

"Because it's a gorgeous piece and so rarely performed or even played on the radio." I smile back just then as the announcer gives the Fantasia as the piece following the commercial break.

When the music comes on, we sit together listening to it; my husband's eyes are closed, I'm watching him. At the final chorus, I take his hand and he slips his arm about me. When it's over I say, "I always wondered how he could write that . . . write music of such joy. This piece is even more baffling to me than the Ninth."

"Why more baffling?"

"Because he wrote this when he knew that it was inevitable . . . when he knew that at some point he wouldn't be able to hear the music anymore. It's that Sisyphean moment when he looked down the hill and knew he would have to slowly make his way to pick up that rock again."

"And you think it should have affected his music?" my husband asks, touching the side of my face.

"How could it not?" I say. "He had to have faced a real and crushing despair then. In *The Mill on the Floss* George Eliot wrote, 'It is in the slow, changed life that follows—in the time when sorrow has become stale and has no longer an emotive intensity that counteracts its pain, in the time when day follows day in dull unexpectant sameness and trial is a dreary routine—it is then that despair threatens.' You see, it's not when you first get the diagnosis or first realize what's happening that you become immobilized by pain. I knew months before they ever said 'lupus' that I had lupus, and in many ways the actual diagnosis was already an anticlimax. No, it's when all the high drama is over and the issue is not whether you are ill, but the fact that you are and won't ever be well again. It's then that you recognize that illness and loss are inevitable and you must know it . . . and live it . . . everyday. It's a grief that—till you face it—is practically unimaginable."

I sit quietly now with my husband as he holds fast to my hand. Though he came into my life much after the fact of my being ill, he has tried albeit as an outsider to understand its history. "But," I say, "in this piece, Beethoven writes not only of beauty and joy as he does in the Ninth, which was years after coming to terms with illness. Here he writes of *Kraft*—at the very time when he must have been completely aware of his own fate. What could be more *dis*empowering than for Beethoven to know he wouldn't be able to hear music?"

"He could still hear music," my husband says. "He would always be able to hear music in his head. He could look at a score and hear the music. All musicians can do that."

"Yes, I know," I reply. "Just as people could read aloud to me from the books in which my own work was printed. It's not the same," I say.

"No," my husband says. "I know it's not the same. But your eyes are better . . . you can read again now . . . your eyes are much better."

"Yes. I've been very lucky in that. But for nine years I didn't *know* that better was possible. And during that time, I could not have written of joy . . . and certainly not of power."

"Which Beethoven did," my husband says.

"Yes. Which he did. But for the longest time I couldn't understand how and I felt that I should. That in fact this was what he was trying to say in this work. I had to listen to this piece over and over and over before I understood how it was that he could sing of power and joy. I finally got it only as I listened to those final four arpeggios and last chords again and again. Here, I'll get the record."

I play the last section of the Fantasia as we listen together once more. I can see that the quality of my husband's attention is quite different now; he is listening not in music as he usually does but in words so he can talk with me about it.

When the piece concludes, I can feel the final chord still hanging staccato in the air between us. "You see, it's there in the music, right there. That it *too* is inevitable . . . each of the arpeggios and the final chords leads to the next. The power comes from that. From the *music's* inevitability. That's what he knew, and that's what he understood."

"But you did write during the years when you were visually impaired . . . you wrote a novel and . . ."

"Yes, but I didn't know *how* I did that until I listened to these chords. I didn't know where that power came from."

"You mean from lifting up the rock the next time?" my husband asks.

"From lifting up that rock the next time," I say.

***Susan L. Feldman** writes fiction and creative nonfiction and is presently working on a short story collection.*

# Weird Science

MICHAEL G. STEPHENS

Science, a nun once told us impressionable young parochial school children, is the devil's workshop. This was not a humble sister, but the principal of my school. I must have been in the fifth grade, still ignorant of even the most basic precepts about science. I did not take her remark with a grain of salt but rather the whole cloth. This was the 1950s on suburban Long Island, a few miles from the great city of New York. The nun's word was ironclad; I spent eight years of grammar school without being exposed to one science class.

We studied geography, social studies, health (the closest we would get to science), and had plenty of math and English. By the time I graduated from eighth grade, I had read Poe, Dickens, and Shakespeare. I knew about the Revolutionary and Civil Wars, about European history, and my sense of mathematics was good, not to mention my penmanship (excellent) and conduct (also excellent). But I never had even a science week or day or hourly chat with a teacher—in those formative years of my education.

The one possible exception would be the time our parents donated our bodies to science. This was done by offering us to Dr. Jonas Salk to test the first polio vaccine, which was administered on Long Island. I suppose that our parents could have said no, but virtually everyone lined up in the basement auditorium of Saint Aidan's grammar school to receive our polio shots. I remember a doctor sitting there with his big needle—this was long before AIDS and everyone received the same needle—and a nurse asked me to look at the bunny rabbit on the wall. I wanted to say, for Chrissakes, I'm not a baby, give me the damn shot, but I looked at the photograph on the wall and then I got stung with the vaccine.

No follow-up procedures attended our vaccinations, no classes or lectures about what had happened to us. We were told that polio would be wiped out thanks to our selfless gestures of donating our bodies to the cause of Dr. Salk. What did I get for my effort? I remember receiving a black and white button which read: POLIO PIONEER. For the rest of that school year I walked around telling myself that I was a polio pioneer, that thanks to my selfless parents, other children would be saved from this crippling disease. I felt proud, though completely unaware of the scientific significance of why I had helped Dr. Salk by getting these early vaccinations.

I carried my scientific ignorance right through high school and into college. Science was the one subject I always did poorly in, the one part of academic life that thwarted any chance of getting a really good grade point average. In my freshman year of high school, I had my first science class. It was biology. I failed. I failed it again in summer school. But why? Superficially, I hated the smell of formaldehyde; spiritually, I had a problem with dissecting frogs, pigs, and rabbits. At the heart of the problem was a total lack of understanding about the scientific method of observation, evidence gathering, and categorization.

I might get A's in English, history, geography, religion, and math, but I kept failing biology until I eventually encountered an easygoing teacher who passed virtually everyone no matter what their grades were. Then, as part of my college prep coursework, I had to move on to chemistry and physics. I did no better. In both classes I did not receive so much as a gentleman's C as I received the gift of a D. I deserved to fail, but each time the teachers said I was a nice kid and had worked hard, so they passed me, but just barely.

If it were not for my poor science grades, I would have graduated near the top of my high school class. All through grade school, in a class of about 150 to 200 students, invariably I was third in my class. Now I was probably fiftieth out of 200 students. The public school I attended on Long Island was excellent academically, highly acclaimed for its science offerings. Grades were very much a component of one's social status, and as a result of my poor showing in the sciences, I did not wind up socially where I thought I deserved to be—drinking milkshakes with the kids who won Westinghouse science awards and were planning to go to MIT, Princeton, Harvard, and Yale —but with a group of boys who didn't care one iota about grades. When I tried to engage the science nerds in conversation, they would pick up their food trays and move to different tables. Still, I had an incredible sense of my own worth as a writing student and as a reader of literature. The science nerds could not or would not engage in discussions about poetry, its metaphors, the charge of emotion, the music in language or Yeats's sense of the grand gesture.

Science was not even on my mind when years later I picked up *The Periodic Table* by Primo Levi. What first attracted me to his work was the fact that he was an Italian Jew. I am originally from Brooklyn, a bumpy seam where Bushwick, East New York, Bedford-Stuyvesant, and Brownsville meet and where the dominant ethnic groups were Italians and Jews. Though I was Irish, I felt more at home—was more welcome—in the Jewish and Italian communities. So I didn't identify science with Primo Levi, but a kinship from Yiddish, the street language of the Jews and the universal tongue of Brooklyn.

Initially I was somewhat baffled by Levi's organizational strategy. Each essay had the name of a chemical, and at first I thought I was going to be reading dry prose about the universal verities of science.

Instead I was given a personal narrative by a scientist, and it was this human element, Levi himself, that drew me into his incredible book. I have not encountered another book quite like it.

Though it is not a scientific treatise, one might learn a great deal about science, particularly chemistry, reading this work. What Primo Levi did was to fool me into liking science as I grew to like his voice. The personal essays that comprise *The Periodic Table* read like a memoir. Yet there is a scientific epiphany in each one. In that sense, nearly every piece is driven by parallel narratives. On the one hand, there is the story of the element whose name graces the title of the essay; on the other hand, we are made to see this element in very personal terms, from Primo Levi's autobiographical point of view.

In twenty-one essays with titles such as "Argon," "Hydrogen," "Zinc," "Iron," "Gold," "Chromium," "Nitrogen," "Uranium," "Vanadium," and "Carbon," Levi uses the chemical property of each as a kind of overriding trope to explode his own story onto the page. These chemicals assume values that one generally associates with images in poetry; that is, they become objects imbued with emotional charge and significance, a kind of scientific objective correlative. Suddenly that which was inhuman takes on human values.

In "Zinc," he has recently become one of the select few to work in the professor's lab. His first experiment is with zinc in an acid solution. At this stage, zinc is simply a chemical, without any metaphoric values: "so tender and delicate zinc, so yielding to acid which gulps it down in a single mouthful, behaves, however, in a very different fashion when it is very pure: then it obstinately resists the attack." Then he observes two conflicting philosophical conclusions: "the praise of purity, which protects from evil like a coat of mail; the praise of impurity, which gives rise to changes, in other words, to life." Levi finds the second observation more congenial. He concludes: "Dissension, diversity, the grain of salt and mustard are needed: Fascism does not want them, forbids them, and that's why you're not a Fascist; it wanted everybody to be the same, and you are not."

Of course, he is not the same, because, like it or not, Primo Levi is a Jew in Italy at a time when being a Jew is dangerous. It did not matter whether his family had been Italians for as long as anyone could remember. Suddenly he had become "the other," the impurity, life itself, but now under a threat of death, and the catalytic impurity that would transform zinc (Italy) if he were not stopped. What makes Levi so amazing is not merely such observations as the above, but rather the way in which he takes a chemical and does not let go of it until he has observed everything about it and himself that may relate to the chemical. In that sense, even if Levi has revealed some of the physical properties of zinc, he is not yet finished with exploring its metaphoric ones. Rita, a woman who also is working with zinc, has caught Levi's eye. While his own zinc reacts with the diluted acid, he takes a stroll around the lab and sees Rita in a parallel world because she also is working with zinc. Levi notes: "at that moment between Rita and myself there was a bridge, a small zinc bridge, fragile but negotiable."

His opportunity comes when he notices that she is reading the same novel by Thomas Mann that he is and that *The Magic Mountain* draws many parallels between the characters in the book and their world in Italy. The discussions between the humanist Settembrini and the Jewish Jesuit Naphtha fascinate Levi. Suddenly, though, Levi is reminded of his own Jewishness. "I am the impurity that makes the zinc react," he observes. "I am the grain of salt or mustard." He notes this because, for months, a Fascist publication had been printing articles about purity. Suddenly, Rita, the lab, zinc, the world around them converge, and Primo Levi thinks about who he is, an Italian Jew, an anomaly, somebody "who should not eat salami but eats it all the same."

Levi's zinc sulfate wound up concentrating, turning to nothing more than a bit of white powder "which in suffocating clouds exhaled all or almost all of its sulfuric acid." With Rita, he had better luck:

> I left it (the zinc) to its fate and asked Rita to let me walk her home. It was dark, and her home was not close by. The goal that I had set myself was objectively modest, but it seemed to me incomparably audacious: I hesitated half of the way and felt on burning coals, and intoxicated myself and her with disjointed, breathless talk.

Then it happens. "Finally, trembling with emotion, I slipped my arm under hers." When she did not resist, he was exhilarated. "It seemed to me that I had won a small but decisive battle against the darkness, the emptiness, and the hostile years that lay ahead."

It was his being Jewish and Italian—and the parallel search for my own roots back in Brooklyn—that first drew me to read Primo Levi. But I have stayed to reread him many times because of the conjunction of literature and science that he brings to the table.

Levi was an Italian Jew, a chemist, who survived a Nazi concentration camp. Yet, to be sure, Levi's book is more human nature than science, more literature than chemistry. In the final essay entitled "Carbon," he writes that his little collection "is not a chemical treatise." For Primo Levi, *The Periodic Table* is a microhistory, "the history of a trade," as he puts it. That trade is chemistry. He goes on to say that he wrote his book at the concluding arc of his career when "art ceases to be long." No mere writer could possibly write of art having a duration; we are taught to believe that art is immortal, not a finite object. And what scientist ever ponders art's longevity in this way? This convergence of writing and science also explains the emotional atmosphere of *The Periodic Table,* the charged particles, the charmed quarks, the eloquent asides, the searing quality of the images that float through the prose like tracers. This suggests that Levi's dual careers are not equal ones, that inevitably the writer consumes the scientist. Would we know Primo Levi at all—no matter how brilliant a chemist—if he were only a scientist, and not a writer? Levi teaches real things about chemical elements in his writing: "Every two hundred years, every atom of carbon that is not congealed in materials by now stable (such as, precisely, limestone, or coal, or diamond, or certain plastics) enters and reenters the cycle of life, through the narrow door of photosynthesis." If he were not such a great writer, I doubt that I would be drawn to these observations. Levi provides me with a high window over which to see the other side, the side where science is interesting, even transforming. Levi brings light to the sciences where once I only had heat.

No matter how great a chemist Levi was, he is remembered primarily as a great writer, perhaps the ultimate survivor, a literary mind who witnessed the inhuman terror, the sheer torture of the concentration camps, and somehow had enough humanity left in him to write about that almost unspeakable—and therefore unwriteable—experience. What does science have to do with Nazis, Italian Fascists, and death camps? Of course, the answer is everything, because one could say that Levi's life was spared because he was a scientist, one who was needed in the factories at the end of World War II. For *The Periodic Table* is nothing if it is not practical, and part of its practicality is to chronicle the life of its creator Levi, as he lived and breathed, not just as a human being, but as a scientist, a chemist, a survivor.

Literature is one door that opens onto humanity, but science is another. Primo Levi was a hyphenated artist, a writer-chemist, and without one I am not sure he could have been the other. I am reminded of the old world of the alchemist, of the ancient scholars attempting to turn dross to gold. It reminds me, too, that there is another kind of alchemy at work here. The alchemy of art, turning the opaque world of science into a scrim through which light is seen. I still imagine a teacher coming up to me at the Bunsen burner and asking, "What in the world are you thinking about now, Stephens?" I might reflect back on Primo Levi and answer, "Phosphorous."

*Michael G. Stephens is the author of* Lost in Seoul *(Random House, 1990), a memoir about living in contemporary Korea; and* Green Dreams: Essays Under the Influence of the Irish *(Georgia, 1994), winner of the Associated Writing Programs' Award in Creative Nonfiction. He is the writer-in-residence at Emerson College.*

# On a Duet, Sung Long After the Music Had Stopped

LAURA S. DISTELHEIM

In every image that I have of them, they are together, John and his mom. Some are remembered images and some are images envisioned from updates I heard, but all of them have that in common: they are together. They were playing checkers, I remember, she on the edge of the bed and he in it, when we pushed open the door of Room 484 that first day. While I settled my nephew, Ethan, in the crib by the window and my sister unpacked his bags, she and John's mother engaged in the "new roommate" ritual, exchanging the profanities that passed for chitchat on that floor: *Chemo,* they said, and *tumor* and *bone marrow aspirate* and *metastases* and *neutropenic,* while folding a stretchie or jumping a king as though saying *grass stain* or *soufflé.*

But then there came a moment when their words careened out of control like a volleyball gone out of bounds.

"What's John's diagnosis?" my sister had asked and, when the answer was "neuroblastoma," the jar of strained peaches she'd been holding slipped out of her hand, splattering in a scream across the floor. That was Ethan's diagnosis. The cancer of infants, his doctors had said—the cancer where detection before the first birthday was the key, where that birthday sliced a line between terror and tragedy. Ethan had been five months old at his diagnosis; John had passed his ninth birthday several weeks before.

"I know, I know," his mother said, forcing a laugh, avoiding his eyes, smoothing the blanket, retrieving the ball. "It's the cancer of babies. Well, this big baby of mine is only the second in the history of the world to be diagnosed at his age." "Here, catch," her eyes begged my sister, "PLEASE CATCH!" and my sister did.

"Wow, John, am I ever impressed," she said. "You're one in a million!" And he grinned and made a muscle with his right arm, the one that was free of IV tubes, before turning back to the checkerboard.

When I came back to visit Ethan the next day, he was riding bareback atop John's knee on the rocking chair, galloping to freedom across a prairie that only they could see. Behind them, near the window, their mothers were talking, looking out at a view which, once witnessed, forever changed every other. "I'll never look at the world the same way again, having been here," John's mother was saying. "I'll never see the same person again when I look in the mirror." The woman who watched her child being wheeled into an operating room threaded with wires and monitors and tubes, she said, just couldn't go back to being the woman who worried that the shoppers coming through her checkout lane would say she wasn't ringing up their groceries fast enough. "Now, I go back to work after spending a week in this place and I tell you, I feel sorry for the first person who complains that I bruised her pears."

"Now you'd just say, 'OH, PULEEZE,' wouldn't you, Mom?" John called from the prairie, and they both laughed.

"We're really pretty lucky, though," she added, trying to laugh again, telling us how her boss let her off every third week to stay with John through his chemo sessions and every time he got so sick in between that he had to be back here again; how neighbors had offered to watch her other two sons after school until her husband got home from work, since the two-hour drive between here and home meant that she couldn't go back and forth. We stood there for a while, I remember, talking and looking out the window, listening to John and Ethan giddyap into the wind, and then my sister and I left for the playroom to gather some toys. As I stepped out into the hall, I looked back through the closing door and, even now, I can still see them, held in the freeze-frame of that glance—John laughing and murmuring to Ethan, his "Bad Hair Year" cap at a jaunty angle and his mother next to his chair, recuffing the sleeve of his robe. The two of them, side by side, piecing together a life raft of ordinary in an ocean of surreal.

The next time Ethan and John were roommates was a month or so later, and the image I gathered then was of the afternoon John and his mom were packing, getting ready to go home. The nurse who'd come in to disconnect his IV had handed his mother a revised chemo schedule, reconfigured just that morning by his doctors. "So it looks like you'll be spending Easter with us," she said, her voice a shade too bright, and I watched John's color slip from pale to gray. "Aww, Mom," he moaned as soon as the nurse had gone, "not Easter . . . ," and his mother began at once to douse the wound with words. But, from across the room, I could see that, beneath the soothing waves of her voice, her arms were cursing.

"Oh well, kiddo," she said, "there's no point in fussing over things we can't change," thrusting his pajamas into his suitcase with a slam-it-all-anyway thud. "There will be other Easters," she said—fastening the zipper with a screech—"and I'm sure the Easter Bunny will find you wherever you are."

But: *Would there?* I heard her arms shout, flinging open the closet, yanking coats from their hooks. *WOULD THERE be other Easters? Was it TOO MUCH TO ASK for him to have this one, this leftover crumb of a disappeared life he was beginning to think he had dreamt?* Obviously, it was. TOO MUCH TO ASK. There was the answer, inked in red, on the schedule she'd been handed: No detours. No delays. No day passes out of this world of elderly children, bald and frail, who'd raced ahead of their parents to peer over life's edge.

"OK, big guy," she said, jamming the checkerboard into a backpack, looking around to make sure she'd gathered everything. "If we hit the road now, we just might beat your brothers home, so let's pack 'em on up and move 'em on out." I held open the door so she could maneuver their luggage cart through it, and then I watched them make their way down the hall, with her arm draped across his shoulders, its fury spent, and his around her waist, ending in a fist that clutched her shirt. The cart's wheels clacked a lament on the floor behind them and every few steps the suitcase shuddered, as if trying to shrug off the emesis basin strapped across its handle, which John would hold on his lap in the car, where Travel Bingo used to go.

Ethan was nearing the end of his course of chemo, with maybe a round or two left to go, when he next shared a room with John. They'd already settled in by the time we arrived, and I remember noticing the weight of the sadness in the room as soon as we stepped inside. A sadness so solid I could almost feel the door bump against it. John had fallen asleep and his mother was sitting by his bed in the rocking chair, unmoving. He'd been losing his hearing, she told us, a side effect of the type of chemo he'd been given, and she and her husband had met with his doctor that morning to tell him they just couldn't let that happen. They'd begged him to change to another type, and when they did, the doctor had said, "Your son is better off deaf than dead."

She shook her head as she repeated the words, still not comprehending them. Wasn't it just a minute ago that it had been asked, "Will your son take a milk shake or a Coke with his burger?" "Will he be trying on dress shoes or sneakers today?" "Would he like a horn or a bell on his bike?" When did the choice get to be between deaf and dead? Ever since his diagnosis, she said, she'd made herself believe there was still a road back to that "before" place—that place of baseballs and tree houses, of skinned knees and fireflies, of sons who squirmed and rolled their eyes when their mothers drew them close. And now this, and so she knew: there wasn't. They'd been evicted from "before" without notice, had spent their last day there and put it behind them without ever having been warned that they'd be leaving. She sat there for a long time after that, watching John sleep, and then she stood up and unpacked a picture of him carving a pumpkin with his father last Halloween, which she placed on the table beside him so that when he woke to find he was still in the "after," he would at least be able to hear a few last echoes of "before."

Ethan had surgery not much later and was then declared "disease free," the tie that had bound him to the hospital now severed, with the exception of checkups once a month. It was when she took him for one of those checkups that my sister again saw John and his mom, there for a test. When his chemo had ended, they told her, he'd had to undergo a bone marrow transplant, but had since been in a remission that had now lasted several months. "You should have seen them," my sister said and, in my mind, I did. In my mind, I saw them smiling, with the lines on their faces softened and their skin bronzed by the sunlight that had flooded back into their world.

I saw them back in their home, their lives now set to a music whose cadence no hearing loss could mute. To the music of a family waking together to a new day, with the newspaper thumping against the front stoop and footsteps padding down the hall, with a shower tapping in the bathroom and a coffee pot hiccuping next to the stove. To the music of a tangle of brothers returning from school, boasting and daring, sprinting and shoving, calling *I'm telling Mom,* as they come through the door. To the music of no more nurses in the night, with fresh chemo bags in tow, of no more doctors huddled outside the door, murmuring over files and notes, of no more meals delivered on whining carts in metal dishes on plastic trays, of no more pain.

I saw John's mother watching him return to flight, become a blur of boyness again as the awful stillness ended—watching from the car as he ran up the school steps, from the porch as he hung upside down from a tree, from the edge of the park as he tore his way around the bases. I saw him pushing the lawn mower next to his father, and I saw her nearby, pulling weeds with his brothers. And then I saw the two of them in the kitchen, he working on his homework at the table and she starting dinner at the stove, both of them all the time knowing these mundane moments as miracles.

I don't know what it was that brought their hiatus to an end, what precise sign it was that told them that the light would soon be leaking from their sky. It may have been a sign as subtle and as shattering as the first leaf to slip from a branch in September, and I imagine that when they noticed that leaf on the ground at their feet, they must have considered taping it back to the tree and pretending they'd never seen it. And I imagine that as more leaves began to carpet the earth around them, they must have wondered what they'd done to invite the winter back.

When my sister told me she'd heard that John had relapsed, she said that because his doctors had run out of artillery, they'd sent him to Amsterdam for a treatment that wasn't available anywhere else. Amsterdam, I remember thinking, to seek refuge in the city where Anne Frank, too, had lived days stained by terror of that knock on the door. When I envisioned them on the flight, on their way across the ocean, I thought of how they must have been thinking that here they were on their way to a place more exotic than they'd ever vacationed in and they wouldn't even see it. And when I envisioned them in the hospital, I knew that even the differentness would be different there, the aloneness painted a starker hue.

And so I envisioned them having the chance to stop off first for some time in a park, to pause for a moment before going to that hospital. But even that image offered no solace, for I saw her sitting erect against the slats of a bench and him limp against her shoulder, watching Amsterdam's children running and climbing, leaping and sliding—and in my mind, I could feel her wondering if there is anything more foreign to the natural world than a ten-year-old boy for whom a jungle gym sings no siren song.

I imagined that as she sat there she was seeing, in her own mind, a day, years before, when she'd arrived to pick John up from nursery school—of how for some reason, on that day, the sight of his jacket and hat and mittens hanging in his cubby and the sound of his high, eager voice through the door had filled her with a need so intense for the world to treat him gently and an awareness so aching that she couldn't force it to do so, that she had felt her eyes fill and had had to turn away from the other mothers who were beginning to arrive. But on that years-ago day, she would now think, she'd been mourning the inevitability of his someday knowing human hurts: knowing the bruise of not being invited to a birthday party, of being the last to be picked for a team, of being turned down by a girl when he asked her to dance. Not this. Not the inhuman. This she hadn't even thought to mourn.

They went back to Amsterdam twice more after that first trip, but the test results after the third time told them that there were no flights to be booked out of this nightmare. The next time I saw them was soon after that final return. I'd gone with my sister to take Ethan for a checkup and we ran into them in the hall. They'd just come from a conference with his doctors and when John went back to search for the hat he'd dropped, his mom told us what they'd said: *Put him in a hospice,* is what they'd said. *It's over,* is what they'd said.

When she was telling us that, as she was saying those words, what suddenly came into my mind was an instant a few weeks before when I'd taken my three-year-old niece with me to a store and she'd slipped from my view. I'd turned to pay for my purchase and when I'd turned back to where I expected to see her, standing just behind me, I found emptiness instead. The whole time John's mother was talking, I was remembering how the absence I'd faced in that moment had been sharper, clearer, bolder than any presence could ever be: an absence written in italics within a frame of jagged edges. I was remembering how, in the moment before I'd found my niece again—stooping to examine a display an aisle away—the space where she had just been, the space where she *should* have been, lowered its lip, stuck out its tongue, flared its nostrils, and taunted me.

I could see from the shadows in John's mother's eyes that when those doctors had said what they'd said, they'd flipped up a window-shade she'd been taking care not to raise, bringing her face to face with the world as it would be beyond that knock on the door—that knock on the door which, in the end, Anne Frank, too, had heard. And I knew that what she was facing was a world in which there'd be no space she could turn to—anywhere—that wouldn't lower its lip and taunt.

But when John came back with the hat he'd recovered, the shadows lifted at once from her eyes. We took Ethan into an examining room then, and they went to the lab to have some blood drawn. When we came out, we saw them for one instant more at the end of the hall—she standing straight, he leaning in, her arm across his shoulders and his around her waist, waiting for the elevator to open, so they could begin their journey home. That's the last image that I have of them, and I know I will carry it with me always: John and his mom, together, staying one step ahead of, one room away from, one day before that knock on the door.

***Laura S. Distelheim** received her J.D. from Harvard Law School. Her essays have appeared in* **An Intricate Weave: Women Write on Girls and Girlhood, Whetstone,** ***and the*** **Chicago Tribune Magazine** ***and are forthcoming in*** **DoubleTake** ***and*** **Pleiades. Grace Notes,** ***her collection of essays in progress, received an award from the Barbara Deming Memorial Fund in 1997.***

# The Lyapunov Exponent

JAMES GLANZ

As we powered up the big electromagnet coils encaging our vacuum chamber one dim afternoon in the lab, R. W. M. warned me never to collect data without a prepared mind, a framework for calibrating gradations of surprise. Since then I must have seen the inside of more Laundromats than laboratories, but the old experimenter's tip on how to sharpen perception when things are about to happen fast has always stayed with me.

Laundromat and laboratory, laboratory and Laundromat: my mentor would understand that unbalanced load of wash just as well as he did the ions whirling in a microscopic spin cycle around the magnetic field lines we dialed into the chamber with a black knob. (Like a harsh detergent, the fields would wipe your credit cards clean if you brought them too close.) The nineteenth-century theorist James Clerk Maxwell based his equations of electromagnetism on not-unrelated mechanical analogies. He filled space with fanciful springs, masses, wires, and fluids in order to imagine the equations that bear his name. By now, those Gedanken cartoons have dissolved and been replaced by the abstract equations themselves as our tales of ultimate reality. The equations nicely describe the flow of light, X rays, radio waves. They also tell how electrical current in our copper coils generated the magnetic fields that, in turn, exerted force on the charged ions. Behind the Plexiglas windows of the vacuum chamber, that force sent the particles into a socks-panties-towel-hankie tumble, a gyrating motion that is also usually expressed with the variables of an equation.

Now, however, I believe those once-benevolent cartoons of existence are returning, like dispossessed spirits, to reclaim what is theirs, to usurp abstractions that in the end have obscured the universe's beauty for all but a small mathematical priesthood. It isn't just a question of Maxwell's equations: a radical degree of abstraction invaded much of science, and without a whimper of resistance. The blankness of those abstractions has gone so far as to crud up the case for scientific research itself—and perhaps to degrade its value. The tidy symbols on a page, even Maxwell's transcendent— $\nabla \times \underline{B} = \frac{4\pi}{c^2}\underline{J} + \frac{1}{c^2}\frac{\partial \underline{E}}{\partial t}$, stand ready to be changed back into expressions of wonder at Nature's workings, into dialog balloons filled with words from Strunk and White's friendly *Elements of Style,* not P. A. M. Dirac's frightening *Principles of Quantum Mechanics.* This resurgence is visible to the careful observer as a pentimento against the black backdrop of empty space.

At the lone pool table of a bar named Kasey's, there is something I would like to tell Vincent, a trader at the Mercantile Exchange who nurtures an unlikely passion for quantum physics and cosmology. The balls click, the smoke from his girlfriend's cigarette stretches into patterns that fluid dynamicists have yet to explain, and the Lares of physical reality seem to dance on the bill of his dirty white baseball cap like Chagall's violinist, jaunty and unmeasurable all at once. For now I listen. Vincent keeps a picture of Albert Einstein in his apartment and sees distant parallels between the theory of relativity and the more provisional laws governing the workings of global financial markets. But within this analogy, Vincent believes, Einstein counts as much more than the equivalent of a trader who bets the house against the British pound and wins. Relativity theory describes *for all time* the distortions of time itself, as when jets carrying ultraprecise clocks race in opposite directions around the globe and show that even the hour at which the markets close depends on one's velocity and acceleration with respect to the Board of Trade.

Does all of this make Einstein a greater historical figure than George Soros, the financier? If so, I wonder, sipping on a pint of golden beer, why Einstein's quip that compound interest was the only miracle he had ever seen? I finally point out close similarities between the two pursuits: Einstein derived his theory by peopling the universe, every speck of it, with imaginary observers who have been given a few simple rules. The speed of light *in vacuo,* for instance, must be the same (call it *c*) for all observers, no matter how they are moving relative to one another. That same *c* slaps a limit on how fast signals can carry information—*Microsoft closed down 12 percent!*—between observers. It develops that such a universe makes sense only if space and time mix together in a curvaceous spacetime described by the deliciously non-Newtonian equations of relativity. Now consider a practical economist, who might assume the world teems with idealized consumers practicing rational self-interest in light of the information available to them at any instant. Whether you base your hunches, your computer models, or your academic papers on those automata, you have created an entire universe one observer at a time; so what's the difference?

Vincent shelters a look of disbelief under the bill of his cap. His father, a tailor in northern New Jersey who grew up outside Naples, built a business on his ability to charm customers. *Misses-a Thomas-a, you lossa few pounds!*—according to Vincent's affectionate caricature. *You looking-a really good.* Vincent cannot understand why an ex-physicist refuses to place relativity theory above the almighty dollar. What I haven't told him, but would like to if there were any opening (in lieu of one I order another Foster's, slouching over a crumpled five dollar bill at the bar), is that I was a truly pathetic physicist. Blessed with scholarships, renowned teachers, and a kind of asphalt-court mathematical quickness, I power-walked past the same Princetonian ivy that caressed Einstein's selfless imagination like cilia within the cavities of a larger intellectual organism. At a trimeter pace that the arches and weathered stone never moderated, I tromped across the outsized lawn in front of the Institute for Advanced Study, where he did some of his greatest work, and I peered *en passant* at the house where his wife would find him forgetfully pondering a problem in the entryway long after he was to have left for his office.

I even shuffled from frame to frame staring at the pictures of lesser luminaries (Richard Feynman, Eugene Wigner) in the corridor between the physics library and the toilets whose stalls were continually marred by suggestive graffiti, such as the constant π written out flawlessly to 212 decimal places. (The New York City area code—*GET IT???*) What obsessed me throughout much of this motion, however, was the Nietzschean metaphor of the labyrinth, in which venturesome thinkers can lose their way and encounter, not a kindly wife, but the devouring Minotaur, far, far from the sympathy or understanding of humans. At other times I would put miles under my feet in a state of terror because I could not refute Nietzsche's assertion that behind every academic work—no matter how intellectually rarefied—one will, after a diligent search, find the author's *intention.*

I could see the business was going badly when, after an absence of several weeks, I returned to the lab with a handwritten translation of a lengthy chapter on matrix algebra from the *Handbuch der Physik* that I thought relevant to our experiments. R. W. M.—a genuinely kind person and one of the few physicists to whom I would listen at all—looked merely worried when I handed him the manuscript. For a time I wrote letters to other scientists I trusted, outlining a growing list of philosophical difficulties and asking whether, like answer keys in the professor's copy of a textbook, standard solutions existed. One dear man, thinking the problem involved isolated matters of faith, reminded me that I was free to ignore any accepted "truths" of physics with which I did not agree. Another warned me that the competitive ambiance of research always rubbed people raw and that my ability to cope would be a measure of my true affection for the subject matter. For several months the high point of my day involved taking lunch in front of the Firestone Library and trying to catch the midday stroll of a man I had been told was John McPhee. Soon I left.

Vincent later corresponds by e-mail, asking me to send him everything I have written on cosmic strings, black holes, Stephen Hawking, and the whole universe of slick-paper science. Improbably, he seems delighted when I comply. But what lodges most firmly in my mind from our meeting is his animated retelling of the *New York Times* obit of Giorgio Strehler, the Italian director. Once again, Vincent noted, here was a nonscientist paying homage to my former *code duello:* The director saw his artistic method as akin to that of a scientist whose experiments lead only to more refined experiments, never to complete truth. Yet, strangely, he did not, in his view, read directly from the book of Nature. He was only an interpreter of a composer's genius rather than a discoverer himself: "Our job is to understand what these great absolute artists have created and communicate that to a public." There is something untoward and obliquely threatening in such modesty, something mocking, especially when one reads it in the *Times,* which my own employer regards as a competitor. Did the man who brought the demons of *Faust* to life truly say this?

Arnold Sommerfeld, the profoundly German mathematical physicist and teacher (1868–1951), wrote in his *Optik:*

> As is well known, Goethe abhorred the theory that white light is a mixture of the seven colors of the rainbow . . . But the rainbow should have convinced him that white light is decomposed into colors by a spectral apparatus (in this case water droplets). In this decomposition the periodicity originates not from the primary sunlight but from the frequency-sensitive spectral apparatus.

Wrote Sommerfeld in another part of the book: "It was the tragedy in the life of Goethe that he would not recognize the distinction between physical and physiological optics; this was the reason for his fruitless fight against Newton."

This is what you think of when you wake in the middle of the night as your cat looks on with a creepy prescience. Sommerfeld's message was that Goethe, like most of us ordinary blockheads, failed to keep the world of his perceptions separate from objective, exterior reality. Since physics applies only to objective reality, the champion of the *Urpflanze* was mistaken in refusing to pick apart the seeming perfection of white light. *("Here I am still, wretched fool!")* The phone rings while the image of my cat, named the Artful Dodger, is still floating in my remembrance the next day. A young-sounding woman asks if I can provide information on an alternative source of energy that comes "from a copper wire," a source that could undermine the oil industry and consequently has been suppressed. I tell her that I could not, but ask what that source is based on.

There is an excited undertone in the voice. "I don't know; I think it's some kind of battery. [Her funders] have been asked by the embassy in Moscow to find alternative energy sources that could be used in Russia. I was told about the copper wire by a retired scientist in California."

"OK. Who was it?"

"I'm not at liberty to say."

"Well, in order to get a handle on what it might be, I would have to know, for example: Is it based on nuclear energy? Chemical energy? Electromagnetic energy? What basic source?"

"Einstein worked on it at the end of his life," she says. "That's all I know." Then her voice changes, like a stream flowing into a bottomless lake: "The scientist in California worked with Einstein."

Visualize two Dutch doors. The doors swing from hinges on opposite sides of a single threshold, like the doors often separating a dining area from a restaurant's or a cafeteria's kitchen. Each Dutch door, however, is split into vertical halves that can in principle swing independently. This leaves four separate demi-doors or panels through which you might pass, say, a tray of bad cafeteria food without moving the other three panels. Call these panels left-up $(\theta_{lu})$, left-down $(\psi_{ld})$, right-up $(\psi_{ru})$ and right-down $(\theta_{rd})$.

In the optics experiment to be considered, the doors are actually polarization states of photons, or particles of light. Imagine that opening a panel corresponds to successfully passing a photon through a polarization filter set to a certain angle. (By rotating a pair of Polaroid sunglasses on a sunny day, you can see the glare off pavement or water brighten and dim at certain angles, since the reflection has slightly polarized the light. A version of the same effect is at work here.) The details are not important, but what happens in the experiment is this: Individual photons emerge from a laser, and then get, in effect, cloned in a special crystal, after which the polarization of one clone is rotated by some amount. The now-distinct photons are each split and the halves reshuffled to make two new photons whose properties are *entangled,* that is, mutually dependent. The entangled photons each strike polarization filters with light detectors behind them.

Let's say the two detectors are shaded by filters set at the unspecified angles $\theta$ and $\psi$. The Cuisinart concoction of the photons ensures that if one detector is shut off, the other may be just as likely to catch a photon passing through its filter as not—no matter what the filter's angle. This means that if a blind person tries to pass a tray through any one of the four panels, she is just as likely to succeed as to bang the tray into a closed door. But if she finds one panel open, the entanglement of the photons can tell her whether any of the others is also likely to be open. In fact, the experimenters (led by Leonard Mandel, University of Rochester, a master of the subtleties of quantum mechanics, which governs what is about to happen) find that if they set the filters at the specific angles $\theta_{lu}$ and $\psi_{ld}$, those two panels always open together. The numerical values, which again don't matter, happen to be 74.3 and

33.2 degrees. The point is that if the blind woman can push the tray she is carrying in one of her hands through the left-up panel, she can always push the tray in her other hand through the left-down panel, as if the two panels were latched together.

*So weit, so gut.* Continuing the experiment, the blind person now finds that the right-hand panels are also latched together, since if one tray gets through $\psi_{ru}$, the other one always passes through $\theta_{rd}$, and vice-versa. She gets down on her knees and finds that something similar is true of the facing panels $\psi_{ld}$ and $\theta_{rd}$: By a mechanism she does not care to explore, they open together at least some of the time. But whatever that mechanism, if the Dutch doors simply have what Einstein, Podolsky, and Rosen (1935) called an "element of reality"—in short, if they behave like honest-to-God doors—then what the blind woman finds next seems physically impossible. She and the experimenters discover that the top panels never open together. Bang! Jell-O, congealing macaroni everywhere. The conclusion, so irritating to Einstein in the context of the cruder experiments of his day, is that there is no contradiction because there exist no doors at all before the blind woman shoves trays at them to "measure" whether the panels are open or closed. It isn't that the photons' polarizations had existed in some unknown but fixed state. *The polarizations did not exist.* Who are we, whose touch transforms possibility into actuality? Can so much depend on our sloppy knowledge?

As a student, R. W. M. had Eugene Wigner for a class on group theory, a highly abstract branch of mathematics that formalized many of the symmetries with which the Hungarian theorist was so obsessed.[1] Scene: A mystified class, furiously scribbling notes, strains to follow as Wigner fills a long blackboard with equations in order to

[1] Given his initials and the other information in this essay, the reader could easily come up with the name. But respect for my former teacher somehow leads me to place him behind the veil of initials.

prove a theorem. He reaches the end of the blackboard, considers for a moment what he has written, and begins erasing it before the desperate students have had time to catch up. "No—I have much better idea," says Wigner.

A sort of modernist adaptation of Wigner's ideas called gauge symmetry now describes the very bedrock of physical reality, according to physicists who sift through the teeniest and most "elementary" particles of which matter is made. Wigner himself is said to have been put off by this development—he would have preferred the universe to spring from classical symmetries in which the elementary stuff looks the same to the left as to the right and the same backward in time as forward. It wasn't to be. Nevertheless Wigner and his groupie theorists knew that whichever symmetry won out, physicists would be grasping for its significance with the tools of quantum mechanics, which provide ways to calculate how that underlying symmetry would manifest itself in an actual measurement. (Usual answer: asymmetrically.) That put him smack up against the Dutch doors whose existence depends on the same measurement. So Wigner thought for decades about how a human action could turn a potential door—symmetric or not—into an actual door.

Eventually he decided there was something about human consciousness that did the trick. "It might seem to the reader that, since I am searching for some kind of role for quantum phenomena in our conscious thinking . . . I should find this view to be a sympathetic possibility," writes Roger Penrose of Wigner's solution in *The Emperor's New Mind*. "However . . . it seems to lead to a very lopsided and disturbing view of the *reality* of the world." Aye, it does, as the philosophy's catch-phrase makes plain: *Consciousness creates reality.* Perhaps because of Wigner's massive prestige, such speculations are not as disreputable among physicists as they were when Sommerfeld confidently announced the tragedy of Goethe's life. This doesn't mean, of course, that they now honor the centuries-long literary exploration of reality as shaped and filtered by human consciousness.

They are like lovable backhoe operators who *will* drive a Caterpillar into the yard to start their first herb garden. Either dill fits into the landscape of the Grand Unified Theory or the cosmos shakes at its foundation.

*Cogito, ergo es.* It does have a pleasantly circular ring. Unfortunately, as so often happens where experimental science is concerned, the philosophical play-by-play soon resembles old film footage of jump shots and broken-field runs that must have seemed remarkable at the time. If the so-far infallible Schrödinger equation holds up, then recently proposed experiments will soon show that merely the *knowability in principle* of aspects of one photon has measurable consequences for other photons caught in a state of entanglement with the first. Consciousness might still be stalking among the precision clamps and polished optics that guide the light; but good luck finding the bigfooted nuisance. My bet is on beasts with fewer unconfirmed sightings and much better conversational skills. Wigner focused more and more on such puzzles as he grew older and upped his involvement with fringy religious groups. Toward the end, like a Julius Erving clinging to basketball for years after his legs had betrayed him, Wigner just had nothing left to give.

One day in Princeton, I caught sight of him sitting in the front row during a talk by Edward Teller. The cold warrior growled at length about the need to maintain our nuclear stockpile against the threats of military strikes and diplomatic deception by the then-Soviets and their allies. "Eugene whispered to me silently and in Hungarian," Teller said at one point, and soon he was discussing the peaceful uses of nuclear energy with Wigner, who had laid many of the conceptual foundations for the first fission power plants. Teller warmed to the general topic of alternative energy sources. As Teller began describing prospects for generating energy with wind turbines, however, Wigner appeared to doze off. Teller stalled, moving to a detailed description of platforms that could be placed far out in the South Atlantic, where the absence of land masses allows the prevailing winds to pile up waves to huge heights. "Those are waves," he chuckled. Giant pistons sticking through the platforms, he said, could be fitted with buoys that would ride the waves and generate electricity. Just as he finished, Wigner's head tottered upright again. "What about energy from the waves?" Wigner asked, to a stunned auditorium. The deeply sympathetic look on Teller's face placed a thought in my mind: *He doesn't really want to nuke anyone.*

Symmetry had won Wigner the Nobel Prize. Around the time of Teller's lecture, R. W. M. walked into his office without the chronic look of worry that political infighting etched on the faces of more than a few of the scientists at the laboratory, where his machine was a bit player. R. W. M. (OK, call him Bob) held an issue of *Science* in which Val Fitch, his own former advisor, described the work behind the Nobel that had just been given to Fitch and James Cronin. They had discovered the violation of a certain kind of symmetry called charge-parity (CP) conservation. Key evidence had turned up in the decay of particles called kaons, which Bob had captured with Fitch long ago, first on mountainsides that are bathed in showers of particles created when cosmic rays scream in from space and smash into the upper atmosphere. Later the two of them switched to kaons made in laboratory particle accelerators. "I reluctantly decided that the future was not in studying cosmic rays in the mountains I loved," wrote Fitch, "but in the accelerators." He thoughtfully mentioned the work with Bob as a precursor to his discovery. "Really gave me a boost," Bob said, glowing.

Poignant fact of Nature: In a roundabout way, CP violation showed that reality is asymmetrical in time, fundamentally different with the clock running forward than backward. Is that asymmetry what lifts reminiscence above the greasy calculations of today and tomorrow? Bob sat in his office, where old announcements for scientific meetings, faded reproductions of his pet paintings, and obscure formulas scrawled on yellowing scraps of paper were taped to the walls, and where some of the bookshelves and filing cabinets could not have taken another reprint inserted by force, and he recalled having relinquished a chance to continue with his particle physics as an entry-level academic. "It sounded like a lot of busy-work," he said, smirking at the notion of grading exams and lecturing to undergraduates. Instead he turned elsewhere, following opportunity and the call of his own brilliance, which lay in making gizmos to chase insubstantial ideas.

"Can I ask you a question?"

This morning I am in a city gym. The baskets are hoisted up, tarpaulins cover the floor and grade-school science projects are arranged in haggard rows. I am a judge, and I am still twenty feet from his project when the boy approaches me with his question. The crown of his head does not reach above my belt.

"Of course," I say to him.

"Do you think a plant can grow in the dark?"

"Well . . . No. I'll say no." As I follow him through the crowd, I see that in front of the cardboard panel explaining his project, there is a shoebox sitting on a table. The box is closed.

"Look at this," he says. He whips the lid off the box.

Inside, there is a yellow, drooping bean seedling.

"Hmm."

"Can I ask you another question?" he asks before I have time to consider the phenomenon of the bean seedling. "Do you think a plant can grow under a desk lamp?"

There is a Styrofoam cup half-filled with parched-looking soil, and otherwise empty, sitting next to the shoebox.

"Yes," I say, "I think a plant should grow under a desk lamp."

"Nope," he says, tilting the cup toward me.

"Did you water it?"

"Watered it, fed it with nutriments, just like the other one. Just like it said in my workbook. I kept it watered good."

I peer into the cup: "Watered it, huh?"

"Three times a day."

It seems to me that he is telling the truth.

"So why do you think it didn't grow?" I ask him.

"I think it had something to do with, when I put the water in, it *sizzled*."

"I think you're right," I say to him. "That did have something to do with it." I toss in a quick suggestion about keeping the desk lamp farther from the soil, then get to the point: "What's interesting about this experiment is that it works just the opposite from what you'd expect. You'd expect a bean plant to grow in the light and not in the dark."

"That's just what I thought!" he shrilled, the gym lights bright in his brown eyes.

This is the haiku of the Laundromat.

> To operate:
> Point arrow at slot
> Insert single coin
> Turn arrow downwards
> Additional coins may be inserted
> for more time.
> —ANONYMOUS

The poem, with its allusions to the unidirectional flow of time and to the free will's action within the constraints of society and Nature, resonates infinitely. But the poem does not tell all there is to be told about the physics of the washing machine. Consider the turbulence of the wash cycle. Clothes, like minds, are either permanent press, delicate, or regular, but whatever the temperature and duration of the wash, those arms, legs, shirttails, brassiere straps, and everted pockets trace the turbulent convection of the fluid in which they are immersed. It is as if these intimate companions of ours were snowflakes carried on the bluster of a storm, making visible the doldrums, bursts, and vortices that we would otherwise fail to perceive. And in that juxtaposition of images one might easily see the same relationship that the Greeks' imaginative sketches of Orion, Taurus, Perseus, Pegasus, and Cassiopeia have to our galaxy's stars as they float among the ionized gases of interstellar space. Indeed, since the stars' proper motions over billions of years will cause their patterns as seen from Earth to evolve, what now seems to be a hunter might one day resemble a peacock, and vice-versa. So it is with the snowflakes that strike us as clothing.

It is a fine experiment to observe pairs of those snowflakes drifting across the gray-white of the sky. Aleksandr Mikhailovich Lyapunov (1857–1918) introduced a measure of turbulence involving the rate at which two bits of a fluid or gas, initially close together, begin drifting apart. In a fluid flowing calmly, without turbulence, those "elements" may drift apart in such a way that the distance between them is linearly proportional to the passage of time. If the flow is turbulent, however, then on average the elements drift apart exponentially fast, at a rate proportional to $e^{\lambda t}$, where t is the time. The constant $\lambda$, the Lyapunov exponent, gives a measure of the turbulence—the larger the value of $\lambda$, the stronger the turbulence. Pairs of snowflakes, although they don't follow every small gust of the wind, become a source of fascination from this point of view.

The snowflakes seem to appear from an infinite height with just this observation in mind. During the quiet, calm, and even sacred moments when the wind is nearly still, the big flakes drift apart reluctantly, as if tethered by an invisible cord. At other times, when the trees shudder against aeolian forces of pressure and drag, two flakes may dance together through one, two, three episodes of turbulent vicissitude before taking their leave with exponential haste. Other pairs begin a ragged flight of separation the moment they are spotted, as tiny as can be, against the uniform brightness of the swarms beyond them.

And when the trees begin waving their boughs, as if imploring the integrodifferential demons mercilessly calculating each eddy of the blast, the snowflakes skitter everywhere, leaping, even cartwheeling apart with a speed the eye can scarcely follow. Strangely, under all conditions, snowflakes are sometimes seen to return for a second embrace, in a manner that appears far from random, and this is a mystery that science must yet resolve in its walks through the snow-covered mountains of Earth and the cosmos.

*JAMES GLANZ received his Ph.D. from Princeton University and is currently a writer at* Science *magazine. His work has appeared in a wide range of publications from* The New York Times *to* Astronomy Magazine.

# A Well-Worn Hanky

SIMON PANG

The driver of the car doesn't speak English very well and his breath smells like rotten fish. He has a small gash on his forehead from breaking the windshield. He says he lost consciousness but I don't believe him. He misunderstands the question. It doesn't matter to me. I'm in no hurry to go to the hospital. He's fine. But if he is sick, if he has a slow bleed inside of his skull and he dies later I will explain to the inquisitors why we stayed so long at the scene. I'll say I told him that we should examine him and take him to the hospital without delay, but he didn't listen. He had to rummage through his glove compartment for his registration. He had to scream into his cellular phone. He was arguing with the highway patrol about towing yards. What was I supposed to do about it? Besides, I like it here on the bridge. We can stay as long as he wants. He can debate with the cop for an hour, I don't care.

When I turn my back on the crashed car, the driver with the bad breath and the puny flares that are like candles hanging on for dear life, I face a deserted bridge. At this hour there is the sound of the wind above all. And below, black water. When I lean over the rail I can see the water moving. There are no cars and because I have never done it before, I run out to the middle of the naked road and touch the abandoned asphalt. I want to feel the grit in my nails. So this is a new experience; I should be happy. Before now I have only known the bridge from sitting in my car, part of an endless parade of traffic. Now I know the bridge for how it truly is: barren. A desolate monument.

I can see for miles down the empty road and there are no cars coming. Nothing of any consequence is occurring. The cop has forms to fill out. He is taking a report from the driver, who is waving his hands in the air and would very much like the cop to understand something. Joe is in the rig. There is no one watching so I dance a little jig. Why not? Right there, for no one to see.

"Did you cancel Fire?" Joe asks a little later.

"Yeah."

"Is he coming?"

"Think so. He says he got knocked out. The cop's talking to him. He can take as long as he wants, I don't care."

I don't like him because he's not hurt very badly. If somebody is hurt very badly and there is great suffering or the loss of life is imminent, then I am charged with helping them. Suddenly, indisputable purpose is thrust at me—a welcome change. It is no longer just a car accident, a fall from a building, a mere tragedy. I become scarcely aware of myself, and I am swallowed up by excitement and the instinct to help the human race.

This guy, on the other hand, busted his car and bumped his head. He has a cut. Ordinary. Here's a good gauge: he doesn't want his car towed because he knows it will cost him a couple hundred to get it out of the lot. What does he think, he can leave the car here? At least he's healthy enough to be concerned with such mundane considerations. I should be overjoyed. Truth is, with all the hype of rushing to a call with lights and sirens, I'm always a trifle put out if the patient isn't primarily concerned with the next breath.

We will take him to the hospital, though. He crashed his car at highway speed, he has a cut on his head from breaking his windshield, and he states he lost consciousness. The protocol considers that significant head trauma and an automatic trip to Highland. Whether he wants to go or not is hardly an issue because I'm not going to allow him to refuse. Too much risk. What if he dies? Too many people know that he thinks he lost consciousness. Would I let someone go who got KO'd after breaking his windshield with his head? I could be in big trouble.

I'm watching him now, gesturing to the cop. He swings his arms wide, he turns his head from side to side, he walks forward, then backward. He certainly doesn't look sick. Finally he comes over with the cop. The cop says a tow truck is coming and he wants to know if we'll be transporting the patient. I nod my head.

"Does your head hurt?" I ask the patient.

"Yes." He says it eagerly, with a smile on his face, and he bobs his head up and down like a marionette.

"Your neck hurts?" I ask again.

Again, "yes." The same uncomprehending smile, the same head bobbing. We walk back to the ambulance and I slap a collar on him and lay him down on the board.

"This is to remind you not to move your head, so you don't sever your spine if you have a broken neck," I say. He has no idea what I said, I can tell. I do the normal stuff—blood pressure, pulse. An IV, for appearances at the hospital. He's worried about his car. He says he needs to get to the restaurant: he's a cook. I say maybe the restaurant won't open today. He is very concerned. I don't care about any of it.

We leave Highland after dropping off the driver with the neck pain. We are posted at Seminary and 580 and Joe is driving. On the way there we stop at the 7-Eleven on MacArthur, to browse. I settle on a cola Slurpee. Afterward we drive to the parking lot of the church on Seminary and park in the darkest corner. After thirty minutes of sitting in the dark and digging at the bottom of my Slurpee I'm jumping out of my skin, though you could never tell. I'm just sitting here, like Joe is, like we all are. I am in the grip of inertia. I am as motionless as a stone Buddha, but inside I'm a flaming wreck. I want to ride the engine of the earth. I want to know how the motor of life works. Take it apart and put it back together again. Instead, I do nothing. I sit here and wait for something to happen.

With the greatest of effort I pull out a book from my bag. *The Confessions of Saint Augustine,* in paperback. The pages are dog-eared and tattered, but I got it that way. Honestly, I can't imagine anyone reading it. I have been determined to get through it, though I am utterly disinterested in what Saint Augustine has to say. Never before has the hunger for spiritual fulfillment seemed so tedious. I've been on the same page for several sittings now. I've carried the book around for a couple of months. There is only one line that I like: "Our heart is restless until it rests in you." I have, of course, no idea what he means, but he sounds so gorgeously optimistic. So sure of the payoff at the end of the desperate, far-flung search.

Dispatch says: auto versus pedestrian. Westbound 580 at High Street, on the on-ramp. We're the closest unit. Of course we're excited. St. Augustine goes flying. Joe lights them up and we go. After two blocks we smell smoke. It's coming from beneath the dashboard.

"Joe, do you smell that?"

"Smells electrical."

"Shit! I can't believe this!"

"Can we make it to Highland?"

"I don't know."

We should call for another unit, but I don't want to. Maybe we can scoop up the victim and make it to Highland before the ambulance gives out. What if we can't? What if the victim dies in the back of the ambulance while we're broken down on the highway? Would that be our fault? We probably shouldn't risk it. We should call for another unit. I sure hate to give up a good call, though. There is no traffic and we're cruising. We're almost there. I turn the last corner and I can see somebody lying on the street, in the crosswalk. The person is still. Looks promising. We're there. Nobody else is. We were so close we beat Fire, we beat police. I wonder who the reporting party was, because there is nobody around. We can hardly wait to get out of the rig. Joe has the door open before we've stopped rolling. Smoke is curling out from beneath the hood. We have to call for another unit.

"Five-o-one, we're on the scene with mechanical problems. One patient seen. Send another unit code three. We'll advise."

The figure still hasn't moved. This is the prospect of discovery: Will the person be breathing, will there be a pulse, broken bones, collapsed lungs, bleeding—what expectation. I shine the spotlight on the figure. It's a man, wearing a dark suit. He is lying on his side, his back toward us and still enough to be dead. We walk eagerly, ready, to his side.

We walk a circle around him, to get a good look. His suit is dark blue with pinstripes. A red tie. His limbs lie in natural angles. Nothing is splay, nothing is skew, not even his tie. Is he breathing? Hard to tell. His head rests on his outstretched arm, as if he were taking a nap. His face is dry and naturally colored. The ground is dry.

"Can you hear me? Are you OK?" Joe asks loudly.

Nothing.

"If you can hear me, I'm going to hold your head still. We're going to check you out and take you to the hospital," I say.

Kneeling down I hold his head in my hands. I look, listen, feel for breathing. He's clean shaven with razor burn. Late forties. His hair, which is shiny with cream and nicely combed, has a touch of gray. I feel no bumps, no soft spots, no blood. There's an inch-long Band-Aid on his chin. I still haven't seen his chest rise but his flesh has life in it. He must be breathing.

"Hey! Open your eyes!" I shout in his ear. Nothing. Joe picks up the man's arm and drops it over the man's head. The arm doesn't drop flaccidly. It hovers in the air for half a second and then swoons to the ground. We frown.

If someone is unconscious they have no muscle tone unless

they're seizing or posturing from a head injury. But this well-dressed man who won't open his eyes or speak has voluntary muscle control. A sure sign of a faker.

What's going on? There is no blood. His clothing isn't torn. I hate him instantly. I pinch him as hard as I can on his shoulder—I have to, after all—to check responsiveness. At first there is no response. Then he squirms a little, then he reaches up with his hand and pulls my fingers off his shoulder. So he localizes pain.

"What's your name?" I demand.

Nothing. He says nothing. His eyes are closed.

"Tell me what happened." Nothing. I pinch him again, harder. I really dig in. Suddenly he grabs my arm again and he raises up on one elbow. So he moves upper extremities and has strong grip strength. I let his head go. His eyes are still closed but I think he's peeking.

"Stop pinching me!" he yells. So he has articulate speech.

"Sorry. Just trying to wake you up. What's your name?"

No response. He drops back down to the ground. What is he doing? What good will it do him to pretend to be unconscious when he just spoke to me? The idiot. Of course, I pinch him again.

"I'm going to pinch you until you open your eyes and answer all of my questions."

The eyes open. So he obeys commands.

"What's your name?"

"Lenny." He mumbles.

"What's your last name?"

"Boxer." The eyes close again, so I pinch him.

"Quit pinching me, goddamn it!"

"How old are you, Lenny Boxer?"

"None of your goddamned business." He brushes me off with his arms.

"We're not going to take you to the hospital unless you tell us how old you are."

"Thirty-nine."

"Thank you, Lenny. Are you hurt? What happened?"

"Hell yes, I'm hurt! I'm hurting all over. I got hit by a car."

I don't believe a word of it, and neither does Joe. I cancel Fire. I've seen people that have been hit by cars before. That eighteen-year-old boy on Bancroft Avenue. Joe and I happened to be just a few blocks away when he got hit. When we arrived there was still a carotid pulse, but he was lifeless. Lying there it didn't look like he was hurt at all—he was as undisturbed as if he were taking a nap on his bed. There were no abrasions, no cuts, no torn clothing. But he was unnaturally still and when I felt his head loose and heavy in my hands, I knew immediately that his neck was broken.

But that doomed teenager was an exception. The usual pedestrian who has been struck by a car displays a panoply of open fractures and pale, sweaty faces. Boardlike abdomens and unstable chests. Blood. The victims can be hysterical or quiet, doesn't matter. One thing I am sure of is that if someone is gravely wounded—or unconscious, like Lenny was trying to make out—his body will plainly admit to it.

Using the pinching technique I discover that Lenny knows what city he's in. He knows the date and the name of the president. No, he hasn't had any alcohol. He doesn't drink nor do drugs, he says.

Joe takes out his scissors and is about to shred the man's trousers; paramedics like to cut off clothing. The rule is that critically injured trauma patients (people who get hit by cars, for instance) have to have their clothing removed—all of it, even socks and underwear. And there usually is no time for buttons. Joe, however, is eager to teach him a lesson. Punitive clothing cutting. Maybe he'll regret concocting this baloney about getting hit by a car if his pinstriped suit is in rags.

"Hey man, don't cut my clothes! Don't cut my clothes!" The panic in his voice startles me.

"You got hit by a car, you could be hurt. Bad. Dying. I've got to see," Joe retorts. He takes one snip with the scissors. The man screams. Even Joe is surprised.

"Please don't cut my clothes, man!"

I signal Joe to hold off. What the hell. The clothes must really mean something to him. I change the subject, to calm him down.

"Hit by a car? What side of your body did the car hit you on?"

"My whole side. My whole body got hit."

"But did the car hit your left side, your right side, your back, front, what?"

"I'm telling you the car hit me!"

"Fine. Where were you standing when you got hit?"

"Right here in the crosswalk. I was walking and I tripped on the piece of fence. I told them about that metal fence. It's a hazard. I called them up about that metal and now I tripped on it and the car hit me because of it."

I try to picture this but I don't see it.

"What fence? Who did you call? What are you talking about?"

"That fence." He rolls onto his side and points to a twisted four-foot section of the highway guardrail that is lying six to seven feet away from us. I have no idea where it came from. There is no damage to the guardrails along the highway ramp around us.

"You mean that guardrail over there?"

"The metal, *the metal!* I tripped on that metal when I was trying to get away from the car."

"You mean the guardrail?" I'm pointing at the guardrail.

"Yeah, that's it, the guardrail. And I'm gonna sue. Somebody has to be responsible for this. I told them. Now I'm gonna sue."

The whole story sounds ludicrous. I don't believe a word of it.

"But did the car actually hit you?"

"Hell yes. I'm gonna sue, too. That piece of metal should never have been there. It's a hazard."

Whatever. Anyway he chooses to tell the story, that's what I'll chart. But I don't believe a word of it.

"Well then, after the car hit you, what happened?" I'm hoping he'll relate a chronological series of postaccident events. That would be an argument that he didn't lose consciousness—more evidence that his claims are bogus.

"What do you mean, what happened?" he says, uncomprehendingly.

"Tell me what you remember next." I speak slowly and measuredly, enunciating every syllable, as if he were a slow child. "What-do-you-re-mem-ber-next!"

"What do you mean, what do I remember next?"

The idiot!

"Well, what happened next?" I spit the words out, losing control of myself. The dunce.

"Then I laid on the street until you guys came."

"How come you didn't get up?"

"Cause I was hit by a car, what do you think? My neck could've been broke, here you are asking me why I didn't get up." Now he's mad. Indignant. But I push on.

"Were you awake the whole time?"

"No, I fell asleep."

I'm furious because I can't get him to reveal what I am sure of, that he didn't lose consciousness. I'm shaking with anger. I want to expose him, force him to admit that he's lying. He didn't get hit by a car. This is fraud, attempted fraud by a half-wit.

Lenny pulls out a tattered, well-worn handkerchief and loudly blows his nose. Then he wipes his nose furiously from side to side, as if he were sanding a knob off a wooden block.

"Excuse me," he croaks. He reminds me of a six-year-old kid who just got knocked down on the playground. I watch him as he stuffs the hanky into his pant's pocket. He mumbles something in between his sniffles. I hear one word: "allergies." Now he takes out a small bottle of Maalox from his coat pocket and takes a quick drink. He mumbles something again. He suddenly seems frail, weak, sick.

"Hey Lenny, you married?"

"Yes, sir."

"What's her name?"

"Alice."

"Don't you think she's worried about you, being out this late?" I ask.

He gives a little chuckle.

"Probably so, probably so . . . she worries about everything."

I picture Alice waiting at home for her husband to return. Does she approve of his mission? Does she know? Is she a fellow collaborator? It doesn't matter. I find myself relieved that this frail and vulnerable man has someone to share his failures with.

I ask Joe to get the backboard and neck collar, to immobilize him. I resume holding Lenny's head still. Who cares if he's making up the whole cockamamie story? At least he's earnest about it. He probably fretted deep into the night for weeks. Maybe even drew a few maps, diagrams, trying to get the scheme perfected. Poor guy. His plan is such a failure and he doesn't even know it yet.

When Joe returns we slip the cervical collar on Lenny's neck and then Joe takes my place holding Lenny's head. I tell Lenny that I have to check him out, from head to toe, to make sure nothing is broken. I tell him that his clothes have to come off.

"Oh no! Don't cut off my clothes, these are the only clothes that I have!"

"These are the only clothes that you have?" I say with disbelief.

"This is my best suit, please don't cut it."

"Well, help us slip your clothes off, then, because if you got hit by a car, they're coming off."

Joe isn't amused; he wants to rip Lenny's clothes apart with his hands. If I were not there he would. I unbutton his dark-blue coat with pinstripes. The polyester fabric is thin and worn. The two front buttons are different shapes and one of them is brown. I begin to unbutton his white dress shirt but he interrupts.

"Don't take my clothes off, it's cold out here."

Yet another demand. The nerve! The man has no idea how close he's come to being unceremoniously stripped, thrown into the back of the ambulance with his head taped to the hard backboard as if he were nailed to it, stabbed with needles the size of pencils, and hustled off to the hospital.

So I surprise myself and send Joe into apoplectic fits when I tell him OK. I will do a cursory check for injuries here and the clothes will come off later, in the ambulance. Quickly I press on his chest: firm, stable. He's wearing a flashy red tie. His white dress shirt is stiff and has creases in it like it just came out of a package; I figure he bought a new shirt for the occasion. Got all dressed up in his Sunday best for his big adventure. Beneath his new shirt is a T-shirt: yellowish, dingy, and frayed along the collar. I press on his abdomen: soft, flat, no grimace. I press on his pelvis and legs: no instability, no incontinence. His trousers, though also a dark blue, are not an exact match to the color of his coat. And the weave of the fabric is off. His brown belt is cracking and torn and a good five inches too big. He's wearing yellowed white socks and brown-and-black Florsheim boots with a half-inch heel that zip up to his ankle. There are scuff marks on the toes and heels that have been carefully touched up with polish. He keeps tugging at the bottom of his coat, straightening out the wrinkles. From twenty feet off he looks quite distinguished, though the ensemble is totally unsuitable for walking around at four o'clock in the morning in East Oakland. Closely inspected I can see his outfit for what it is: a collection of odds, ends, and hand-me-downs. I certainly don't see what all the fuss is about harming his clothes.

Another ambulance comes roaring up to us. It's Keith and Darren, two cowboys. They love talking in loud voices about saving people.

"Hey Simon, whatcha got?"

Keith comes over, all swagger and optimism. I tell Keith and Darren what's going on. I tell them that the patient states he tripped on the guardrail and then got hit by a car. Keith and Darren are looking at Lenny with a skeptical eye. They see no blood, no torn clothing, no grotesque angles. Their minds are racing, and I don't want them to think I don't share their suspicions. I say that as of yet I have found no injuries. I go further. I say that the patient's story is inconsistent. Then I really stick it to Lenny: I say that the patient seems eager to pursue litigation over the guardrail being negligently left on the roadside. I say it because I don't want Keith and Darren to think that I didn't see right through the ruse, that I'm not sufficiently cynical, macho, competent, indignant. Keith and Darren are foaming at the mouth, eager to tear into the liar.

Now I try to pull back. I mumble something about Lenny being a nice guy. I want to say that Lenny has a wife named Alice who is waiting for him at home, worried. That he carries a hanky in his pocket. But they are deaf now. Liar, faker, 911 abuser, waster of their time, that's what's roaring in their ears. They hear nothing else. Keith has his scissors out, he's clipping them together like a crazed barber. Keith is going to punish this man for wasting his time, for mocking his lifesaving skills. I want to say that it doesn't matter who Lenny is or what he has done, we can be nice to him if we choose. I know what they would say, that somebody could be dying a block away and we wouldn't be able to respond because we were tied up taking care of crap like this. But the truth is there are no other calls. The city is quiet. The only person for us to help is Lenny. I want to tell them that he doesn't deserve meanness. Are we that disappointed that he's not mangled and bleeding all over the street?

Too late. Keith has cut Lenny's trousers to shreds. I am strangely silent. I'm surprised he did it so fast. Now the shirt and coat. Ribbons. Lenny has his hands over his face. He is crying.

***Simon Pang is a writer living in Berkeley, California. He also works as a paramedic in Oakland and San Francisco.***

# Blur (The Interior of a Diagnosis)

RUTH GILA BERGER

The first days Zeezel's paw patter is quick and desperate. It ends with an elongated meow and a befuddled cat face turned up to me. The sound expands, fuzzing the air. My throat is ripped by the air's newly grown thorns, and I concentrate to make each breath very small and slow. She wants something from me. Voices loom over me sucking me into the mess of a long abandoned knitting-yarn basket. They all belong to family members, schoolboys, shrill, shriek, shout, hiss, rant. The tarry shadow they create finally homogenizes into a dull roar. *Nothing stupid not real whore slut nothing nothing inhuman incapable of anything failure selfish nonfeeling whore nothing stupid slut dirty Jew nothing nothing.* I claw my way through this loud mud, squinting my eyes to reshape the noise to its point of origin below two brown triangles I usually know to scratch. A few days ago this creature was clearly a cat. A cat who followed me, bumping around my legs. A very long Siamese who chatted up at me from my lap while I wrote. Now she wants something from me. I cast around frantically for the old boot of common sense. I know Zeezel's searching meows are simply a response to an unusual stimulus, my presence on the couch at twelve in the afternoon. The feline alarm has sounded as if to say, "Mom's home! She's not in her office. She can play!" As my fingers graze an ear my body curls, shrink-wrapped in fatigue.

• The National Institute of Mental Health reports that during any one-year period 17.6 million American adults experience a depressive illness. Depressive illness clusters into three different kinds of disorder: bipolar disorder, dysthymia, and major depression (chronic or episodic). A diagnosis of major depression is precluded by the following conditions: physical illness, alcohol, medication, street drug use, normal bereavement, bipolar disorder or mood-incongruent psychosis (for example, schizoaffective disorder, schizophrenia, schizophreniform disorder, delusional disorder, or psychotic disorder not otherwise specified). Chronic major depression is diagnosed in 20 percent of people who fall into the depressive illness envelope.

Days run by tangling themselves into weeks. My husband waits like a sunning tortoise to find a moment where my attention catches fire. "Have you called your therapist?" I shake my head. I don't like her, she's too maternal, I say. I hate therapists, my mom is a therapist, I say. Craig looks at me, "That's not true, you liked Kathrine." That was two years ago. Kathrine was offered a job working with chronic schizophrenic adults in a group home setting. The job would facilitate her licensure so she left the clinic where I saw her. Hey, I said, I'd dump me too. Depressed woman versus a guy who thinks he's a fourteenth-century Christian martyr? No contest.

Craig laughs through turned-down lips. He pulls back the quilt to touch me. His hands dance over my skin before settling to knead my rocky shoulders. When his fox-red haired fingers skid into something sticky Craig yanks the covers away from me. I have many tiny beaded scratches all over my right side at the waist. Oh, I say, the cat. Before turning a circle and tucking her head into her feet she had been kneading me into an acceptable surface. Her furry wrists flexed with sharp little hooks. I don't know if I noticed at the time, I say. I don't tell him about the newly clotting scabs on my chest, how my skin skimmed off like unfeeling butter and stuck under my frantic fingernails. The blood stickiness was reassurance tasting of salt, proof I still existed. Craig's steady blue eyes narrow slightly in my gaze. I fall away from their distinct blue into the blue lid of the bottle of bleach sulking on its shelf by the washing machine. My tongue grows into a word-soggy sponge forcing me to swallow and regain my focus. "Are you OK?" He asks. I nod, hoping to crack the ice that covers my brain and prevents thoughts from reaching their destination. His hands flex and gel around mine forcing melted wax to transform into a sitting human form. "You're crying," he whispers, skimming tears off my cheek with tender, careful plant-fingers. An angry whale slams its tail inside my body. I look to where my locked-together legs touch. Knobs laced with a map of my childhood injuries shake at the end of my thighs. Craig touches the scar that resulted from my jumping out of a moving Jeep in junior high. "You're shaking," he says. The whale hits my gut low and a few wet drips spread in my underwear, bringing consciousness to me. I, I, I, ha-hav-v-e to—pee, I stutter, smaller than a mouse before I slump into Craig's chest. My last tears barely find a home in his shirt before I'm all warm, dull, and dark. "Sweetie, sweetie, let's get you to the bathroom." He runs fingers through my hair, sifting through dust stuck in two-and-a-half weeks' grime until my eyes hit a clock that registers twenty minutes have gone by since either of us spoke. Two rubber forms previously dusting the floor become my feet and we creak up the stairs. I concentrate on the grip of my spreading toes on each stair. With every twitch of quadriceps and hamstring the roar winds up so it is a tornado in reverse, smashing my organs and structure against my skin. Hollow-bodied I watch myself reach the turn of steps before the second floor. *Nothing stupid not real whore slut nothing nothing inhuman incapable of anything failure selfish nonfeeling whore nothing stupid slut dirty Jew nothing nothing.* The bottom of the stairs croons all cushiony and sensual as Play-Doh to me. Nothing would be harder than the hammers and howls slamming around my head. I jerk from Craig's grip in a downward motion but he catches my fall, forcing me to sit. My face falls numb to the glittering tears that fill our silence.

- The lifetime risk for major depressive disorder is 10 to 25 percent for women and from 5 to 12 percent for men. At any point in time, 5 to 9 percent of women and 2 to 3 percent of men suffer from this disorder. Prevalence is unrelated to ethnicity, education, income, or marital status. The average age at onset is twenty-five, but this disorder may begin at any age. Stress appears to play a prominent role in triggering the first one or two episodes of this disorder but is not always necessary in its subsequent ones.

The blind glass lake seeping toward Craig and my feet is not a new one.

*Blue flowers blue flowers. I have blue flowers on my underwear. Blue flowers are special because most flowers are pink or yellow and girlie. I'm not girlie, I'm blue, hard as the sky. I can fly and nobody knows except my cat Whitering. She has a magic, perfect, orange, black, and white striped tail. On the day I was three we had a tea party and she told me I could only fly until I turn five. We put the stars in our pockets and dropped them to make a trail home. We know what the trees are thinking when it rains. Now that I don't wet the bed anymore I can wear my blue-flowered underwear every day. If I spin very very very fast after I get dressed I'll become a hurricane, gigantic and invisible. All my bones are secret. I have a hole in my heart that has to be fixed with surgery but Ray my grandma's boyfriend is magic. At their place in the country night shadows dance ballets. The screen on my bedroom window cuts the moonshine up into a billion tiny pieces that crash across the floor until it's even too late for bird songs. Ray has bluer than fancy marble eyes that dart into my room away from my grandma in her daisy lace slip I sometimes play dress-up in. He sits on my bed and pets my hair tracing its long tangles down to my back, loud fly-buzzing breaths slip out his mouth and stomp around the room when his fingers touch into my butt. The bed springs creak, he pulls my underwear with craggy nails, moves, and then the world hurts and I can't breath. The world is falling dark. I'm dead.*

I can't do this, I say. "Do what?" Craig asks. Live. My answer cracks me up. When did I get so damn melodramatic? This is so stupid, I say, picturing myself sauntering in a pink feather boa. I pretend to throw it over my left shoulder. Craig stares at me. "But you can, I've seen you get through this before. Tell me what pulled you out the last time." He shakes me but I am a winter tree. Nothing falls. "We'll get through this." I wish I could sweep the grit from his voice, put it in a collecting bottle. I know my pain affects him, little barbed stones he swallows. My wishes to cook us a meal to soothe his throat, shrimp in a garlicky tomato cream sauce, go dancing or curl up and veg in front of the TV, or discuss the politics of Star Trek are all irrelevant. I sit still.

- The average length of a depressive episode is approximately nine months. The course is variable. Some people have isolated episodes that are separated by many years, whereas others have clusters of episodes, and still others have increasingly frequent episodes as they grow older. About 20 percent of individuals with this disorder have a chronic course. The risk of recurrence is about 70 percent at a five-year follow-up and at least 80 percent at an eight-year follow-up. The number of previous episodes is an important risk factor for recurrence.

My cheek melts into the cool, white heartbeat of the bathroom tile. Cobwebs dangle from the bottom basin of a steel globe above me. "Ruth." I jerk. "Watch your head on the sink." The curve swims closer.

"Goddammit!" Craig squats. "Ruth. This is enough. Honey, you're here, you're an adult who is safe now." His voice slices my head from between deflated red kick-balls that bounce off each. Blah, blah, blah. After my head stops wobbling I recognize the balls I saw are his lips. They continue to move. "Ruth, I need you to call your shrink." I turn into a hornet and buzz through my teeth at him. "Ruth . . ." Craig's voice fades into pieces of soggy debris left over from past episodes. I wonder which reel he spins. You're planning to leave me, the whine jumps out of my head before I can grab it. This isn't going away, I say.

The first time Craig saw me come out of a major depressive episode, we went for a walk. Holding hands against the crisp, fall apple-tart air and dancing leaves we decided to walk around the mansion-lined streets by the lake. After checking out all his architectural favorites we got into our usual argument about how no one needs that much money or house. It took us ten minutes to realize we were on the same side. I tweaked Craig's nose and took off running until he caught me and we landed laughing in a leaf pile, heaving puffy white breaths out into the darkening air. "Did you ever have to rake leaves when you were a kid?" He asked. Heading to the grocery store we traded stories. "I wish I had your memory. . . ." Craig's words stopped. "But not the content," his whisper finished. On the way home he said he wished he knew more about depression, watching his feet crunch stray yellow leaves. But he was afraid that reading up would make it permanent and rocky-cliff real.

"Ruth . . ." Each breath I take shoots out a different picture of myself, all ugly cockroaches that should be smashed. "OK. Fine. Why don't I run you a hot bath. I'll hold your hand so nothing bad will happen and when you get out we'll call the doctor." Water fights and loses its voice to the roar in my head. The floor beckons with a magnetic comfort. The fingers I use to touch it are wee and scared, the same ones that fingered ripped, blue-flowered underwear. I crawl into a corner where my shoulders feel like enormous bird wings that can protect me. Leave me alone, I tell the hovering shadow. He retreats and returns.

• Poor outcome or chronicity in major depression is associated with the following:

- inadequate treatment
- severe initial symptoms
- early age of onset
- greater number of previous episodes
- only partial recovery after one year
- having other severe mental disorder (for example, alcohol dependency, cocaine dependency)
- severe chronic medical illness
- family dysfunction

Metallic bleats invade the next mush of days. The fact I know this sound as the telephone is knowledge that blasts its pupa fight out of an almost fossilized cocoon. Each time it rings I burrow nose-first deeper into the planet of quilts, but I still shake. Each hang-up bangs around my ears. My shrink calls back and leaves an angry-sounding ocean of concern on the machine. His psychiatric nurse calls twice. The floor between me and the phone shimmers with the dangerous heat of an unknown entity. Its uneven brown rolls like waves above a hell maw or clouds above a canyon. Either way my toes freeze into cold steel. They refuse to let me set them on it. Chair legs and strewn shoes morph to hazard-size with giant voices echoing *Nothing stupid not real whore slut nothing nothing inhuman incapable of anything failure selfish nonfeeling whore nothing stupid slut dirty Jew nothing nothing.* Craig has removed all the knives and razors from their proper places in the house. Sleep crawls into my eyes before I can rummage around to find them. When I excavate sight from under crusty eyelids my thoughts snap, all Venus flytraps with one response to the voices. Nah-

nah-na-nah-na, I still have one razor. It's twirled with wire into a sculpture I made entitled "Abused Woman." Nah-nah-na-nah-na! I'm trickier, smarter, and better than you. I'm better than you, Craig, better with all my black-nail thoughts. Giggles bubble out of my mouth as I gurgle and glow, curled around my new secret, happy for the first time in days.

- Up to 15 percent of patients with severe major depressive disorder die by suicide. Over the age of fifty-five, there is a fourfold increase in this death rate.

At bedtime Craig falls asleep before I do. I slither to the bathroom and am electric. Naked, I spin in the mirror trying to distinguish myself from my genetic influences, from my mother, who explodes out of me. With terrible eyes I inspect every inch for the undepressed me.

The real me. The one who has a herd of soapstone elephants on her writing desk, who prints out a page that screams YUK in large letters over and over when she can't write. The woman who thrilled at the fact her husband wore her garter belt to work. The woman who volunteered for a rape crisis line and did a presentation on how to handle problem callers. She holds court just behind my unseeable shadows. The visible ones hold my mother. I've got to reestablish my boundaries. Got to locate my essential differences. I sit on the edge of the toilet and pick up tweezers, my faithful lover. *Nothing stupid not real whore slut nothing nothing inhuman incapable of anything failure selfish nonfeeling whore nothing stupid slut dirty Jew nothing nothing.* My mother had a giant tangle of pubic chains. Big stupid purple genitals she stroked in the bath with me smiling that I should do the same with the little kid ones I had. I pluck hair by hair. Each one catches and slides out with a sound I imagine like the clear clinking of champagne glasses. Each hair is a sword pulled from under a stone. Blades to chop my mother away. I turn over these rocks to find myself and reel from

10:24 to 4:56. A bird's nest of curled black hairs floats on top of the water. These piled tiny wires tenuously cover the voices softening their volume. Little red blood polka dots are not purple. Craig stumbles to find me all icicle. "What are you doing?" I disappear the tweezers into my palm, ever eager with my magic tricks. "Come back to bed." His hands pry me up and pull stiff-knit bones back to our lair. Dreams cyclone me to the dank muddy gray of my parent's basement. I wake up but the image doesn't even waver.

Memory has crept up and sucker-punched me to attention. Age eleven with my Yankees baseball cap, diary, and softball mitt I sit in front of a white, metal cabinet laced with abstract flowers of rust. A few minutes of jiggling a bobby pin in the lock and the doors open, pulled apart from their seal of accumulated crud to reveal household toxics. Mr. Clean, rust remover, bleach, paint thinner, turpentine, epoxy glue, silver polish, Lestoil, machine oil, windshield wiper fluid, engine oil, assorted smudgy jars of unidentified muck and radiator fluid. I uncap the radiator fluid and absorb its alien glow. Mr. Yuck, Mr. Yuck, Mr. Yuck my stomach makes its silent chant. Mr. Yuck. An hour unravels into minutes into buzzing seconds as I pour the green into a plain pink Dixie Cup and breathe slow, a ritual toast before consumption. My fingers shake through the miles from the floor to my chin. My tongue lashes tears of frustration off my satiny cheeks. I don't have the guts to drink. *Nothing stupid not real whore slut nothing nothing inhuman incapable of anything failure selfish nonfeeling whore nothing stupid slut dirty Jew nothing nothing.* Can't even succeed at suicide, too afraid to die, I put the cup down, then take it and spill its contents down a quiet, lonely drain in the far corner of the floor before curling up in a laundry basket to sleep.

My adult legs twitch, kick as I fall through many years of pretzeled attempts, the two boxes of generic allergy pills I threw up, the vomited quantities of aspirin, the expired Valium I stole from someone's mother's dental prescription, the never-enough lines to make an overdose of coke, the blind-drunk drives home on thin, mountain-squiggle roads.

- Suicidal risk can be assessed by whether or not certain factors are present.

    - Is the individual feeling depressed or hopeless?
    - Is the individual having suicidal thoughts?
    - Is there a history of previous attempts?
    - Is a concrete plan for a suicide in place?
    - What is the planned method? (Guns, asphyxiation by hanging, or carbon monoxide are more lethal than most sleeping pills unless combined with alcohol.)
    - Does the individual have the means immediately available?
    - Has the individual been under extreme stress or experienced a recent loss?
    - History of chemical dependency, social isolation, loss, mental or physical illness, and age.
    - Has the individual been setting their affairs in order, saying good-bye to loved ones?

I've got a razor, I've got a razor. I have no pubic hair, my legs whisper together smooth, and my left armpit is completely clean. Ha ha. Craig has gone to work after programming crisis numbers into the phone, after making me sign a contract to call while pointy, starred, silent laughs flew out of my lips, nasty little birds. My fingers were crossed I think before freezing. The couch the couch. I don't move. A day and a half later the cat jumps on my chest to nuzzle under my chin. I pet her and call Craig at work. Baby I need you, help me. I'm scared. Worries about hospitalization fall to the floor floating on drafts like used tissues. I voice them. I don't want to go to the hospital, I want to stay home with you and the cats. Heroic mama-bird on his bike he returns, all shirt wings flapping to bring me food and the phone. We call the shrink. He dials a number, hands me the receiver, and settles to hold me.

- For patients with severe major depressive disorder, 76 percent on antidepressant therapy recover, whereas only 18 percent on placebo recover. For these severely depressed patients, significantly more recover on antidepressant therapy than on most modes of psychotherapy. The course that is thought to be most effective is a combination of both antidepressants and psychotherapy.

We hang up, Craig's hand slides off mine like wind-blown sugar as I put the receiver down. "Are you going to the hospital?" I shake my head and tell Craig where the last razor is so he can hide it. The shrink is aggressively upping the dosage of the antidepressant I am currently on, Effexor. "To make you an effective citizen," Craig jokes with a mama-bear snarl.

Now I play the wallflower, the doctor has done his jig, made his move. Will this new color of pill put the switch back in my hips? Allow me to waltz away from his office and back into the sun-speckled robin's egg that is my life? Three years ago Zoloft proved ineffective even at the maximum dose. Wellbutrine followed. On it at the maximum dose I remembered that I loved watching winter mornings knit together their silky appearance. Bad days came out of their corners sometimes, but essentially I was well. Well on Well-butrine, I quip. Craig groans.

We knit our words to create a cohesive thought. Craig holds a flashlight into the recent past. He pauses to wipe the cobwebs that obscure our view, then cleans his hands on velvety, gray-sky-colored jeans. He remembers how I called my psychiatrist in a panic a few weeks before my parents were to arrive for a visit. *Nothing stupid not real whore slut nothing nothing inhuman incapable of anything failure selfish nonfeeling whore nothing stupid slut dirty Jew nothing nothing.* My monsters were starting to stomp and howl, I feared I was slipping into an episode. In an attempt to head it off the doctor switched my medication despite all his little rat-man notes that I need to be on near-toxic levels of medication. Because it's not safe to alter too many neurotransmitters at once, I had to be weaned from Wellbutrine first.

My parents came and spit their thunderstorm. My father sat with his eyes scripting a neon sign that proclaimed me crazy. Sitting straight with his pride-laced spine he claimed superior fatherly awareness. He knew what Ray did to me. The fact he didn't intervene is an irrelevant popped balloon in his reality. Hail raged from his eyes, biting my cheeks as I grabbed for impersonal neutral topics and my mother eyed me like a choice drumstick, waiting to jump and kiss her slobber onto my neck. The thunderstorm streaked its fire through my fingers and left acid in cold pools just behind my eyes.

My parents left just as I was starting Effexor at a minimum dose. Before Craig or I could mop up the flood, I slipped under its black. Now I sit beside little piles of pills, each one bigger than the last. Craig laughs when I say Snap, Crackle, and Pop have moved in just behind my ears. "The side effects will fade," he says, echoing the pharmacy sheet. I wait, toes tucked under me, knowing my breath by the cigarette smoke it expels.

*Ruth Gila Berger is a Minneapolis writer working on her MFA in creative writing at Hamline University, where recently she served on the 1998 editorial board of the* Hamline Journal. *In 1997 she was named a finalist by S.A.S.E. for the Jerome Fellowship. "Blur" is her first published essay.*